Critical Maths for Innovative Societies

THE ROLE OF METACOGNITIVE PEDAGOGIES

Zemira Mevarech and Bracha Kramarski

BETTER POLICIES FOR BETTER LIVES

This work is published the responsibility of the Secretary-General of the OECD. The opinions expressed and arguments employed herein do not necessarily reflect the official views of the OECD member countries.

This document and any map included herein are without prejudice to the status of or sovereignty over any territory, to the delimitation of international frontiers and boundaries and to the name of any territory, city or area.

Please cite this publication as:
Mevarech, Z. and B. Kramarski (2014), *Critical Maths for Innovative Societies: The Role of Metacognitive Pedagogies*, OECD Publishing. *http://dx.doi.org/10.1787/9789264223561-en*

ISBN: 9789264211384 (print)
ISBN: 9789264223561 (PDF)

Add Series: Educational Research and Innovation
ISSN: 2076-9679 (online)
ISSN: 2076-9660 (print)
DOI:10.1787/20769679

The statistical data for Israel are supplied by and under the responsibility of the relevant Israeli authorities. The use of such data by the OECD is without prejudice to the status of the Golan Heights, East Jerusalem and Israeli settlements in the West Bank under the terms of international law.

Photo credits:
© Aakash Nihalani ("Sum Times")

Corrigenda to OECD publications may be found on line at: *www.oecd.org/publishing/corrigenda*.
© OECD 2014

Foreword

As scientists and engineers drive so much of our innovation and creation of knowledge, high-quality science, technology, engineering, and mathematics education is key to the success of advanced economies. Given its transversal nature, mathematics education is a cornerstone of this agenda.

Beyond nurturing the talent of mathematicians, scientists and engineers, good mathematics education can also foster the innovative capacities of the entire student population, including creative skills, critical thinking, communication, team work and self-confidence.

This book explores how to achieve these goals. Based on a review of state-of-the-art experimental and quasi-experimental research, it argues that new types of problems should be featured in mathematics curricula, and shows that pedagogies that emphasise metacognition have an impact on mathematics outcomes, including mathematical reasoning, communication and math anxiety, from kindergarten to university level.

Among the many findings of the book, two have especially caught my attention. First, pedagogies that highlight metacognition are even more effective in collaborative settings. Second, their effectiveness is enhanced when they address both the "cognitive" and "emotional" dimensions of learning. Singapore has pioneered the large-scale adoption of this approach, explicitly emphasising metacognition in its maths curriculum. Interestingly, it is also one of the top performers in mathematics and problem solving in OECD Programme for International Student Assessment (PISA). That suggests that changes in pedagogy could better prepare students to develop the kind of mathematical skills that they will need in more innovative societies.

A complement to two recent books from the Centre for Educational Research and Innovation (CERI), *The Nature of Learning* and *Art for Art's Sake?*, this book is designed to assist practitioners, curriculum developers and policy makers alike in preparing today's students for tomorrow's world.

Andreas Schleicher
Director for Education and Skills

Acknowledgements

This book was authored by Zemira Mevarech and Bracha Kramarski, Professor and Associate Professor, respectively, at Bar-Ilan University (Israel). Drawing extensively on their research on metacognitive instruction, Mevarech and Kramarski designed the book in close collaboration with Stéphan Vincent-Lancrin, Senior Analyst and Project Leader at the OECD Centre for Educational Research and Innovation (CERI), Directorate for Education and Skills.

Within the OECD Secretariat, the book was edited by Vincent-Lancrin and Carlos Gonzalez-Sancho, Analyst (CERI). Francesco Avvisati, Kiira Kärkkäinen and Gwénaël Jacotin made helpful comments on the first draft of the book. Anne-Lise Prigent commented on a later version of the book. Sally Hinchcliffe copy edited the book; Jacotin prepared the graphs; Riitta Carré and Rhodia Diallo formatted the manuscript and provided assistance throughout the process; and Lynda Hawe and Anne-Lise Prigent coordinated the final publication process. Last but not least, Dirk Van Damme, Head of the Innovation and Measuring Progress division in the Directorate for Education and Skills, is thankfully acknowledged for his continuous support throughout the project.

The book is an output of the CERI Innovation Strategy for Education and Training, a project led by Vincent-Lancrin. One strand of the project explores how curricula and pedagogies can better equip students with "skills for innovation", that is, technical skills, skills in thinking and creativity, and social and behavioural skills. It is often believed that artists, scientists and entrepreneurs are key actors in countries' innovation systems. The project therefore synthesises the evidence about the impact on innovation skills of arts education, science, technology, engineering and mathematics (STEM) education, and entrepreneurship education.

Critical Maths for Innovative Societies: The Role of Metacognitive Pedagogies thus complements a previous book of the project: *Art for Art's Sake? The Impact of Arts Education*, as well as recent reports from other CERI projects, notably *The Nature of Learning*.

Table of contents

Boxes

Tables

Figures

Follow OECD Publications on:

 http://twitter.com/OECD_Pubs

 http://www.facebook.com/OECDPublications

 http://www.linkedin.com/groups/OECD-Publications-4645871

 http://www.youtube.com/oecdilibrary

 http://www.oecd.org/oecddirect/

This book has...

StatLink 🔗

A service that delivers Excel® files from the printed page!

Look for the *StatLinks* 🔗 at the bottom of the tables or graphs in this book. To download the matching Excel® spreadsheet, just type the link into your Internet browser, starting with the *http://dx.doi.org* prefix, or click on the link from the e-book edition.

Acronyms and abbreviations

ALN	Asynchronous learning networks
APTS-E	Aptitude Profile of Spatial-Visual Reasoning Test
BASC	Behaviour Assessment System for Children
CAS	Computer algebra system
CASEL	Collaboration for Academic, Social and Emotional Learning
CHT	Collaborative Hypothesis Tool
CUN	Complex, unfamiliar and non-routine
DCD	Developmental co-ordination disorder
EL	E-learning
ERL	Externally regulated learning
ES	Effect size
F2F	Face-to-face
FOK	Feelings of knowledge
GPS	Global positioning systems
ICT	Information and communications technology
IST	Intelligent tutoring software
JOL	Judgement of learning
KMA	Knowledge-monitoring accuracy
KMB	Knowledge-monitoring bias
MAI	Metacognitive awareness inventory
ME	Meta-experience
MMT	Multilevel metacognitive training

MRT	Mental Rotation Test
MSLQ	Motivated Strategies for Learning Questionnaire
NCTM	National Council of Teachers of Mathematics
NJMCF	New Jersey Mathematics Curriculum Framework
PCK	Pedagogical content knowledge
PLC	Professional learning communities
PISA	OECD Programme for International Student Assessment
RA	Reflection Assistant
RIDE	Respect, intelligent collaboration, deciding together, encouraging
RULER	Recognise, understand, label, express, regulate
RS	Reflective support
SEL	Social-emotional learning
SMS	Short Message Service
SRL	Self-regulated learning
SRQ	Self-regulated questioning
STAD	Student Team-Achievement Divisions
TAI	Team-Assisted Individualization
TGT	Teams-Games-Tournament

Executive summary

Education has changed dramatically over recent years from elite education provided to only a small percentage of the population to compulsory education in which no child should be left behind. The skills necessary for the industrial age have been superseded by those deemed appropriate for the knowledge-based world. Our models of learning have also progressed: instead of seeing learners as *"tabulae rasae"* ("blank slate") simply absorbing information, we view them as active builders of information, constructing knowledge.

The dramatic changes in our understanding of the nature of learning have resulted in shifting the focus from the "what" to the "how". There is a broad consensus that in innovation-driven societies, teaching basic mathematics skills is necessary but insufficient. Schools have to guide students in solving complex, unfamiliar and non-routine (CUN) tasks, and foster greater mathematical creativity and better mathematics communication. This approach is for example reflected in the OECD Programme for International Student Assessment (PISA), adopted by more than 65 countries and economies as of 2014. Yet the "million dollar question" still remains: how to enhance students' abilities to solve both routine and CUN tasks.

Education researchers have examined how such tasks are executed. A wealth of research has indicated that metacognition – thinking about and regulating thinking – is the "engine" that starts, regulates and evaluates the cognitive processes. On the basis of these findings, various models have been developed to help students regulate their behaviour during the learning of mathematics. Among them are those developed by Polya, Schoenfeld, and Verschaffel, IMPROVE, developed by Mevarech and Kramarski, and the Singapore mathematics curriculum. All these models provide techniques for training students to use some form or another of self-directed metacognitive questioning in maths problem solving. These models work best in a co-operative learning environment where students study in small groups, articulate their mathematical reasoning and describe their heuristics. In all of them, the teacher plays an important role in explicitly modelling the use of metacognition.

Metacognitive pedagogies have been largely examined in the educational research arena. Among these methods, IMPROVE is the one which has been most widely studied. Being rooted in solid social-cognitive theories, IMPROVE has established itself as an evidence-based practical method. In IMPROVE, self-directed questions act as a scaffold for comprehension thinking ("What is the problem all about?"), connections thinking ("Have I solved problems like that before?"), strategic thinking ("What strategies are appropriate for solving the task?") and reflection thinking ("Am I stuck, why? What additional information do I need? Can I solve the problem differently?"). These self-directed metacognitive questions are generic and therefore could be easily modified to be used in other domains, such as science, reading, or even for fostering social-emotional outcomes. Research findings show that this metacognitive pedagogy is:

- Effective across all educational levels: in kindergarten, primary and secondary schools, and in higher education.

- Usable for CUN and routine tasks, although the effects are notably more apparent on CUN than routine tasks.

- Easily modified for use in other domains (e.g. science), since the self-addressed metacognitive questioning are generic.

- Platform-free, meaning it can be embedded in various learning environments, including co-operative learning or Information and Communications Technology (ICT).

Studies have shown that IMPROVE students outperformed their counterparts in the control groups on routine "textbook" problems, CUN tasks and authentic problems. These positive effects were found in arithmetic, algebra and geometry. Moreover, IMPROVE showed lasting effects even in high-stakes situations such as matriculation exams. Generally, metacognitive pedagogy positively affects also lower achievers, but not at the expense of the higher achievers. Recent studies have shown that IMPROVE has positive effects also on science literacy.

Current research in neurosciences has shown how cognitive and emotional systems are intertwined in the brain. Thus, improving children's social-emotional skills can have an impact on their learning. Modified versions of IMPROVE and other similar metacognitive pedagogies can therefore be used to improve not just academic achievement but also affective outcomes such as reduced anxiety or improved motivation.

Observations have shown that many teachers use metacognitive processes in their teaching implicitly, but they rarely explicitly teach metacognitive skills. Professional teacher development programmes have started to include some elements of metacognition. Limited studies have found that implementing metacognitive pedagogies into professional development courses appears to be effective at improving teacher's knowledge and skills, and their judgement of how likely they are to put what they have learnt into practice. Yet, none of these studies have followed the teachers into their classrooms to assess the impact on their teaching or their students.

As the field of metacognition in general and metacognitive pedagogies in particular continues to develop, there needs to be links between research, practice and policy. Given that most countries aim to adopt evidence-based policy making, the strong evidence regarding metacognitive pedagogies should result in greater focus on these approaches. Teachers, principals and policy makers do not have to reinvent the wheel in their attempts to implement these methods. The principles are well known and have been delineated in many studies. Metacognitive pedagogies are one way for mathematics education to prepare students for innovative societies.

Key findings and recommendations

- CUN problems should be one pillar of all mathematics education in innovation-driven societies, not just for gifted students.

- Solving such problems requires students to implement metacognitive skills, particularly regulating their thinking through planning, monitoring, control, and reflection.

- Metacognition can be taught in "regular" classrooms with ordinary teachers. Improving metacognitive skills has positive benefits for academic achievement, particularly for CUN problem solving.

- Students need to be explicitly taught how to activate these processes and given ample opportunity to practice.

- Given the importance of domain-specific metacognition, learning environments should embed metacognition in the learned content.

- The self-directed questioning used by most metacognitive pedagogies is an effective way to teach metacognitive skills and can be adapted for use with students at any age and in different disciplines.

- These methods are most effective when combined with co-operative learning environments, and applied across more than one discipline.

- Metacognitive pedagogies can also be used effectively within ICT-based learning environments, including asynchronous learning networks, cognitive tools, mobile learning and domain-specific software.

- Metacognitive guidance can be used to improve social-emotional outcomes such as reducing anxiety or increasing motivation.

- Metacognitive pedagogies that combine both cognitive and motivational components seem to be more effective than those applying only one component.

- Teacher professional development programmes should be restructured to emphasise metacognitive processes over content and skills, although more research is needed into how effective teacher training translates into practice.

- International joint efforts could greatly advance research and development of metacognitive pedagogies aiming at improving schooling outcomes in innovation driven societies.

Introduction

This book is based on tens of studies, all trying to understand how education can foster the skills that are appropriate for innovative societies. It focuses on mathematics education, a subject that is heavily emphasised worldwide, but nevertheless still considered to be a stumbling block for many students. While there is almost a consensus that the mathematics problems appropriate for the 21st century have to be complex, unfamiliar and non-routine (CUN), most of the textbooks still include only routine problems based on the application of ready-made algorithms. The challenge might become even greater as the development of mathematics literacy comes to be one of the key aims in the curriculum. Undoubtedly, there is a need to introduce innovative instructional methods for enhancing mathematics education and in particular students' ability to solve CUN tasks. These require the application of metacognitive processes, such as planning, monitoring, control, and reflection. It will be critical to train students to "think about their thinking" during learning.

In the following pages we explore the questions:

- What types of mathematics problems and sets of skills are useful in innovative-driven societies?

- What are the higher-order thinking and metacognitive processes that will enhance learners' ability to solve CUN and routine mathematics tasks?

- Which metacognitive pedagogical models have been developed for these purposes?

- What is the evidence to supporting the metacognitive approach? Are the effects of these methods beneficial for different skills simultaneously, or is there any kind of trade-off?

- To what extent are the effects of the various metacognitive pedagogies evident in school and post-secondary students?

- How could the different kinds of metacognitive scaffolding be embedded in ICT environments to help the improvement of mathematics education?

- What are the implications for pre-and in-service professional teacher development?

These basic questions are key factors for decision makers, principals, teachers, parents, educators and researchers. By answering them, a picture of how to implement effective mathematics education will emerge. For example, recognising that it is not sufficient to teach students to apply ready-made algorithms to maths problem solving may result in accepting the importance of training them to apply metacognition learning. Recognising that young children in kindergartens and preschools could also benefit from metacognitive training might lead to a different approach for early education. Knowing that metacognitive guidance can be supported by information and communication technology (ICT) environments might encourage teachers to use technology in mathematics classrooms. The good news is that all these powerful processes can be successfully taught in ordinary classrooms at all school levels and in higher education with or without ICT.

This volume is based largely on studies into mathematics education, but also includes a few examples from science education. It has focused only on "typical" students in ordinary schools, although there are plenty of other studies that examine the effects of metacognitive instruction on children with learning disabilities or special needs, and other groups.

The first chapter focuses on the types of mathematics problems and sets of skills that are useful in innovation-driven societies. It describes CUN tasks, mathematics reasoning, creativity, problem posing and communication.

The second chapter describes the higher-order metacognitive thinking processes that enable people to solve problems. It reviews models of metacognition, including those of Flavell, Brown and Schraw et al. It also discusses the differences between cognition and metacognition, general and domain-specific metacognition, and the debates about the development of metacognition as a function of age.

The third chapter provides a general overview of metacognitive instruction starting with the question of whether metacognition can be taught. It asks "what is the role of co-operative learning in facilitating metacognitive and cognitive processes"? Do these processes need to be explicitly practised? And finally, what are the key elements of metacognitive pedagogies?

The fourth chapter focuses on metacognitive pedagogical methods in mathematics education. It describes models developed and implemented in various countries, including those developed by Polya, Schoenfeld, Mevarech and Kramarski (called IMPROVE), Verschaffel, and the Singapore mathematics curriculum. The chapter concludes by analysing the similarities and differences between these models.

The fifth chapter reviews the evidence supporting the use of IMPROVE and similar metacognitive pedagogical models. It reviews the immediate, delayed and lasting effects of these models on mathematics achievement in routine, CUN and authentic tasks across all age groups, over different periods of time, in different areas of mathematics and in high-stakes situations. It also examines the preferred conditions needed for implementing these programmes.

The sixth chapter analyses the effects of metacognitive pedagogies on social-emotional outcomes such as reducing anxiety or increasing motivation. It reviews three kinds of metacognitive interventions: 1) those focusing on cognition and metacognition but not explicitly on emotional processes, on the assumption that enhancing cognitive-metacognitive outcomes will also improve emotional factors; 2) those focusing on improving emotional factors and through that attempting to increase cognitive achievement as well; and 3) the combined approach, focusing on both cognition-metacognition and emotional outcomes, on the assumption that both are needed. In all these studies the effects of the metacognitive pedagogies were compared with "traditional" learning with no metacognitive interventions.

The seventh chapter focuses on embedding metacognitive guidance into educational environments which have been enhanced with the use of information and communications technology (ICT). It describes three kinds of ICT environments: 1) specific mathematics software; 2) e-learning, including asynchronous learning networks and mobile learning (short message service or SMS); and 3) general technologies adopted for mathematics education, such as mathematics e-books. It compares the effects of these technologies with and without the support of metacognitive guidance.

The eighth chapter discusses applying metacognitive interventions to teachers' pre-and in-service professional development programmes. Participants in these courses play the double role of being students and teachers simultaneously, with implications for the design of the interventions. It also evaluates the use of combined ICT and metacognitive guidance for pre-service teachers. Finally, it indicates how teachers judge their learning via metacognitive instruction, and the extent to which their judgment of learning is accurate compared to "traditional" learning with no metacognitive guidance.

Finally, the last chapter discusses the implications of these pedagogical models for mathematics education.

The studies reviewed here are only appetisers. This volume aims to provide a useful knowledge base for understanding what metacognitive pedagogies are all about, how to implement them in the classrooms, in what way they enhance routine and CUN problem solving along with social-emotional outcomes, and their benefits and pitfalls. These studies provide evidence on the course of action recommended for developing quantitative literate citizens that can contribute to and thrive in innovative societies.

Chapter 1

Mathematics education and problem-solving skills in innovative societies

Problem solving is at the core of all mathematics education. The solution of complex, unfamiliar and non-routine (CUN) problems has to be the cornerstone of any effective learning environment for mathematics for the 21st century. While students solving routine problems can rely on memorisation, solving CUN problems requires mathematical skills that include not just logic and deduction but also intuition, number sense and inference. Innovative societies require creativity in mathematics as well as in other domains. The approach to mathematical communication has also changed, with students in all age groups being encouraged to engage in mathematical discourse and share ideas and solutions as well as explaining their own thinking. Developing these competencies may result in enhancing social skills as well as mathematically literate citizens.

Complex, unfamiliar and non-routine problem solving

Mathematics is taught in schools at all levels, four to five times a week. Undoubtedly, most mathematics schoolwork involves problem solving. As Stanie and Kilpatric point out in their review on "Historical Perspectives on Problem Solving in the Mathematics Curriculum", "problems have occupied a central place in the school mathematics curriculum since antiquity... The term "problem solving" has become a slogan encompassing different view of what education is, of what schooling is, of what mathematics is, and of why we should teach mathematics in general and problem solving in particular" (1989, p.1, cited in Schoenfeld, 1992).

Although problem solving in mathematics has been taught from the time of the Greeks, if not before, the concept of problem solving has changed dramatically in the last decade. In the past, "problem solving" has referred mainly to the application of ready-made algorithms to the solution of routine exercises and word problems. Yet, according to the OECD Programme for International Student Assessment (PISA), the assessment of mathematics skills for the 21st century should focus on the "capacity of students to analyse, reason and communicate effectively as they pose, solve and interpret mathematical problems in a variety of situations involving quantitative, spatial, probabilistic or other mathematical concepts" (OECD, 2004, p.37). Students have to be "mathematically literate" – they have to "possess mathematical knowledge and understanding, apply the knowledge and skills in key mathematical areas ... and activate their mathematical competencies to solve problems they encounter in life" (OECD, 2004, p.37; see also OECD, 2013).

The term "problem solving" has two components: the type of problem to be solved, and the knowledge and skills needed to solve the problem. The traditional type of mathematical problem includes arithmetic computations, certain equations, geometry problems and "routine" word problems that usually consist of two or three sentences that include the mathematics information, and a question that guides the students in constructing the appropriate equation to solve the problem. In geometry, students are presented with the properties of shapes and theorems for proofs (OECD, 2004). Usually, all the information needed is given in the problem, and the students are asked to apply the theorems in what has to be proven.

Clearly, the skills needed to solve these types of problems are limited, and teaching these skills usually consists of demonstrating the appropriate technique followed by a series of similar problems for practice (e.g. Schoenfeld, 1992). In spite of the fact that the development of mathematical thinking is one of the main objectives of mathematics education, Yan and Lianghuo (2006) found that most of the problems in mathematics textbooks were these kinds of routine problems, where it is usually obvious what mathematics is required. As a result, many students admit that memorisation is the most important skill they need to succeed in mathematics classrooms (Schoenfeld, 1992).

In contrast to these traditional mathematics problems, the type of mathematics tasks suitable for the 21st century differs not only in the content, construct and contexts in which the problems are posed, but also in the processes needed to solve the problems. According to PISA (OECD, 2004, 2013, 2014), the *content* brings up the mathematical big ideas, the *context* often relates to authentic real-life situations ranging from personal to public and scientific situations, and the *constructs* are more complex than in traditional problems. Problems may include mathematical information that is not always presented in an explicit form, and may also have multiple correct answers. These problems for tomorrow's world may consist of a full paragraph of text in which the mathematics information is embedded. Students are asked to make decisions based on their mathematical knowledge and the processes they carried out. Quite often, the problems include different kinds of representations, and sometimes also require students to search for additional information either using computers or other sources. Computational problems may also differ from the traditional ones in asking students not only to carry out the computations but also to explain their reasoning and how they solved it. Often, students are asked to solve the given problem in different ways, to suggest creative solution processes, and to reflect on and criticise their own solution and that of others. These types of problem solving are typical of the PISA 2012 problem solving component (OECD, 2014). This is not to say that routine exercises and problems are to be excluded from the curriculum. On the contrary, routine problem solving is necessary for practising, attaining mastery and being able to respond automatically. But mathematics education has to go beyond routine problems to include innovative problems that are complex, unfamiliar and non-routine (CUN).

Another characteristic of mathematics problems suitable for the 21st century is that there could be multiple correct solutions. Innovative problems such as those described above are authentic and presented in real-life contexts that often pose questions to which there is more than one correct answer. The solution of problems which may have multiple correct answers depends on the basic assumptions that the solver adopts. On the basis of these assumptions, the solver constructs a flowchart with multiple routes. Working in groups may expose the solver to other sets of assumptions for which there are different solutions, and/or different strategies for solution. Under these circumstances, it is essential for learners to reflect on the outcome and the processes used.

Box 1.1 provides three examples from the same context (buying and selling). The first is a very open task to which there are multiple correct solutions depending on the set of assumptions and information the students select to solve the problem. The second, the Pizza task, is a more open task with specific information embedded in the problem. The third is a "routine" task.

Naturally, these different types of problems require different kinds of processes and skills for solution. Given that the problems are based in a real-world context, the students have first to identify what the problem is all about and what mathematical knowledge has to be activated in order to solve it. To do so, students have to bridge between their existing knowledge and the information provided in the task. Then, progressively, solvers have to suggest strategies for "transforming the problem into one that is amenable to direct mathematical solution" (OECD, 2004). The final steps involve some form of reflection on the outcome and its completeness and applicability to the original problem (OECD, 2004, 2013, 2014).

Box 1.1. Examples of CUN, authentic and routine tasks

The supermarket task – an example of a CUN task:

Before the holiday, several supermarkets advertised that they are the cheapest supermarket in town. Please collect information and find out which of the advertisements is correct.

The pizza task – an example of an authentic task

Your classmates organise a party. The school will provide the soft drinks. Your task is to order the pizzas. The class budget is NIS 85.00. Of course, you want to buy as many pizzas as you can. Here are the menus of three local pizza restaurants. Please compare the prices and suggest the cheapest offer to the class treasurer. You have to write a report to the treasurer in which you justify your suggestion.

	Price per pizza	Diameter	Price for supplement
PIZZA BOOM			
Personal pizza	NIS 3.50	15 cm	NIS 4.00
Small	3.50	15	4.00
Medium	6.50	23	7.75
Large	12.50	38	14.45
Extra Large	15.50	45	17.75
SUPER PIZZA			
Small	8.65	30	9.95
Medium	9.65	35	10.95
Large	11.65	40	12.95
MC PIZZA			
Small	6.95	25	1.00
Large	9.95	35	1.25

A sale – An example of a routine task:

In supermarket A, 1 kg of meat costs EUR 8 and 1 kg of poultry costs EUR 4. In supermarket B, 1 kg of meat costs EUR 7 and 1 kg of poultry costs EUR 5. Mr Jonson wants to buy 3 kg of meat and 2 kg of poultry.

Which supermarket is cheaper?

In the context of PISA, the various competencies required for employing these processes are specified as follows: "thinking and reasoning, argumentation, communication, modeling, problem posing and solving, representation, and using symbolic, formal and technical language and operations" (OECD, 2004, p. 40).

In summary, new types of mathematics problems that are complex, unfamiliar, and non-routine (CUN) and go beyond traditional problem solving are likely to be better adapted to preparing students for an authentic use of mathematics. These types of problems refer to formal as well as to real-life situations, involve co-ordination of previous knowledge and experiences, include various representations and patterns of inferences, have one or multiple correct solutions, and prompt reflection on all stages of the problem solving. While the solution of CUN problems is based on "traditional" knowledge and skills, they also require additional, higher-order skill sets.

Mathematical reasoning

Reason refers to the capacity to make sense of things, to establish and verify facts, and to change or justify practices, institutions and beliefs. Mathematical reasoning includes proofs, logic, cause-and-effect, deductive thinking, inductive thinking and formal inference. Thus, mathematical reasoning is based on the ability to reflect on the solution, apply judgment and be able to articulate one's own mathematical thinking. Quite often, mathematics reasoning includes also intuition, number sense and inferences that are both rigorous and suggestive (Steen, 1999), even though formal proofs are perhaps more frequently evident for professional or advanced mathematicians than for mathematics students.

Enhancing mathematical reasoning is an integral part of primary and secondary school standards as described by the US National Council of Teachers of Mathematics (NCTM) (2000) and others such as the New Jersey Mathematics Curriculum Framework (New Jersey Mathematics Coalition and the New Jersey Department of Education, 1996). It is also in line with the PISA definition of mathematics literacy (OECD, 2004, 2012). For example, the New Jersey Mathematics Curriculum Framework (NJMCF) declares:

All students will develop reasoning ability and will become self-reliant, independent mathematical thinkers... Mathematical reasoning is the critical skill that enables a student to make use of all other mathematical skills. With the development of mathematical reasoning, students recognize that mathematics makes sense and can be understood. They learn how to evaluate situations, select problem solving strategies, draw logical conclusions, develop and describe solutions, and recognize how those solutions can be applied. Mathematical reasoners are able to reflect on solutions to problems and determine whether or not they make sense. They appreciate the pervasive use and power of reasoning as a part of mathematics...Students must be able to judge for themselves the accuracy of their answers; they must be able to apply mathematical reasoning skills in other subject areas and in their daily lives. They must recognize that mathematical reasoning can be used in many different situations to help them make choices and reach decision. (New Jersey Mathematics Coalition and the New Jersey Department of Education, 1996, p. 1).

The NJMCF standards summarise the importance of mathematical reasoning, calling it "the glue that binds together all other mathematical skills" (New Jersey Mathematics Coalition and the New Jersey Department of Education, 1996; Resnik, 1987).

While there is broad consensus regarding the need to enhance mathematics reasoning in primary and secondary school classrooms, there is still much debate over what it means. Sometimes it refers to formal mathematics based on the use of exact mathematical language. Other times it denotes intuitions, insights, sense making and informal inferences described in less rigorous language (Steen, 1999). For some teachers, the use of informal mathematical language contradicts the very essence of mathematics. For others, doing mathematics is not limited to formal proofs. Steen indicates that formal mathematical reasoning is useful for solving the routine problems presented in textbooks, but not necessarily for solving what we call CUN problems for which "formal reasoning is only one among many tools" (Steen, 1999, p.1). For example, many ICT programs guide students to draw inferences based on searches of poor-quality data, either including more information than necessary or incomplete or imperfect information. The solutions to authentic problems are also often based on intuition and heuristics, and thus cannot be limited only to "formal reasoning". According to this approach (Steen, 1999), it would be inappropriate to base the teaching of mathematics only on formal reasoning, in spite of its importance. Formal reasoning is just one kind of mathematics skill.

Mathematical creativity, divergent thinking and posing problems

Mathematics is clearly about problem solving, which is often associated with the "technical skills" of the field, the "knowing-how". Yet in innovative societies it is fundamentally important to be able to think "out of the box": create original ideas and construct connections between different objects, approaches or disciplines.

Although the topic of mathematical creativity is a "critical one" (Sheffield, 2013, p. 159) in innovation-driven societies, and although it has been emphasised by the OECD (2004, 2014) and many other organisations, it has been largely neglected in the mathematics education research arena (Leikin and Pitta-Pantazi, 2013). Several inter-related reasons may explain this neglect. First, there is no agreed definition of creativity in mathematics. Second, there are almost no tools to assess mathematical creativity. Finally, little is known at present how to develop mathematical creativity in schools.

Creativity is usually conceptualised as a form of divergent thinking involving the generation of multiple answers to a given problem (Guilford, 1967). This is in contrast to convergent thinking aiming towards a single, correct solution to a problem. Activities that promote divergent thinking include constructing a set of questions, brainstorming or designing mathematics games. Torrance (1966) defined creativity as "a process of becoming sensitive to problems, deficiencies, gaps in knowledge, missing elements, disharmonies, and so on; identifying the difficulty; searching for solutions, making guesses, of formulating hypotheses about the deficiencies; testing and retesting them' and finally communicating the results" (p. 6). Torrance's Tests of Creative Thinking identified four main components of creativity: fluency, flexibility, originality and elaboration (Torrance, 1966). Fluency refers to the total number of meaningful and relevant ideas generated in response to a stimulus; flexibility is the shift in approaches taken when generating responses to a stimulus; originality is the statistical rarity of the responses; and elaboration includes the amount of detail used in the responses.

About two decades later, Sternberg and Davidson (1995) identified three components in the mental processes associated with creativity: 1) use of different representations; 2) constructing mental connections between different objects and providing explanations and justifications; and 3) solving problems of different kinds (in Leikin and Pitta Pantazi, 2013). Sheffield (2013) proposed a five-stage non-linear model of creativity: relate, investigate, communicate, evaluate, and create. According to her observations problem solvers may start at any point (component) and proceed in any order, often repeating several processes as the problems are more clearly defined, explore possible solution, and pose new questions. They can make connections between the problems under consideration and previous mathematical knowledge, use a variety of strategies to investigate possible solution, create a variety of solution, models and related questions, evaluate their work throughout the problem-solving process and not just the end, and communicate with peers, teachers, and other interested adults while working on the problem as well as flowing its solution (Sheffield, 2013, p. 326).

According to Wallas (1926) the creative process involves preparation, incubation, illumination and verification. All the models emphasise the importance of relating to prior knowledge.

Although these studies analysed creativity without referring to any specific domain, these models could easily be applied to mathematics education. Mathematics teachers can train their students to solve problems in different ways; shift between arithmetic, algebra, geometry etc. and ask students to provide original solutions. For example, solving the Supermarket task (Box 1.1) is often based on divergent thinking, requiring the solver to be flexible, fluent and original, as well as being able to provide elaborated responses.

Under the umbrella of mathematical creativity, we sometimes include also problem finding or problem posing, referring to a large range of competencies from formulating a question or questions related to a given mathematics text, through to the discovery of innovative problems.

In mathematics classrooms, finding problems is usually used as a means to facilitate problem solving. During the teaching of solving a particular type of problem, students also practice finding problems relating to the type of problem introduced. In finding/ posing problems, proposing a large number of problems represents "fluency", shifting between different kinds of problems is considered "flexibility", and suggesting unusual problems indicates "originality". However, finding problems can be much more difficult than solving the problem. The classical case of Fermat's Last Theorem exemplifies a situation in which the identification of the problem was clear, but its formal solution took three and a half centuries (the theorem was published in 1637 and its proof in 1995).

The very definition of problem finding relates to creativity which also involves the search for alternative solutions, identification of innovative problems, or reconsidering the problem's definitions (Silver, 1997; Csikszentmihalyi, 1996).

The development of student problem finding has been largely experienced in mathematics classrooms (for an excellent review see Silver, 1997). Fluency and flexibility are often fostered by asking students to generate multiple correct solutions to open-ended, ill-structured problems, or by guiding students to propose different strategies for solving a given problem. Discussing and evaluating the solutions presented by classmates may guide students to provide multiple solutions (fluency), look for different kinds of methods (flexibility), and suggest novel solutions (originality). The pizza task presented in Box 1.1 exemplifies how creativity can be fostered in the mathematics classroom. It calls for various interpretations in order to complete the task. The students have to decide which items to include in the analyses, the quantity of each of the items, and the prices to include (regular or on sale). Similar processes are also applied in the solution of the supermarket task in Box 1.1. A student who decided to only include food products in the analysis might obtain a different answer than one who included a sample of all items or those items that are sold more frequently. Clearly, these problems do not have a single correct answer, and the different answers depend on the basic interpretation of the problems. These kinds of problems can be administered at different educational levels, depending on the mathematical knowledge and competencies of the students.

In sum, "creative thinking is a cognitive activity that results in finding solutions to a novel problem. Critical thinking accompanies creative thinking and is employed to evaluate possible solutions" (OECD, 2012; p.13). CUN problems can help to develop some of those dimensions of divergent thinking and creativity. They also enable students to deal with uncertainty and decision making.

Mathematical communication

Communication in mathematics refers to reading, writing and talking about mathematics. Sometimes all three competencies are pulled together under the umbrella of mathematical discourse.

The basic approach to communications in mathematics classrooms has changed during the late 1990s and beginning of the 2000s. Before then, teachers believed that their main role was to disseminate knowledge, facts and algorithms, and generally expected students to replicate them (Brooks and Brooks, 1993); communication in mathematics classrooms was mainly carried out by the teacher. He or she is the one who "talks" mathematics, introduces the new mathematical concepts to the students by using "mathematical language", and explains the mathematical symbols and terms to be used in solving the problems. Under these circumstances, most teachers relied heavily on textbooks (Ben-Peretz, 1990) and students worked individually to master the new procedures and algorithms. Since the textbooks largely include routine problems (Yan and Lianghuo, 2006), and since students work individually on these problems, there was little room to encourage students to discuss, explain, or be involved in any kind of mathematical discourse.

Furthermore, some teachers oppose the idea of involving students in mathematical communications because: 1) most of the students who have difficulty with mathematics also have also difficulty with reading and writing, and therefore the teachers are concerned that emphasising communication would impose more difficulties on the students; 2) the mathematics curriculum is heavy and intense leaving no time for emphasising skills other than "pure" mathematics; 3) reading and writing are part of the humanities rather than mathematics courses; and 4) giving students the opportunities to "talk" mathematics may result in them using terms imprecisely, whereas mathematical language should be exact.

The change in the content, processes and context of mathematics education for the 21st century has naturally led (or should naturally lead) to an alteration in the basic approach to mathematics communication in the classroom. When students are confronted with CUN problems rather than with procedural algorithms, sharing ideas, discussing solutions and explaining one's own thinking is unavoidable. Whether communicating in writing or orally, students have to be clear, convincing and precise. Light and Mevarech (1992) indicate that mutual reasoning is an effective means for achieving cognitive change because giving explanations and listening to others' ideas provides students with an opportunity to look at the solution in different ways and reflect not only on their own solution but also on those of others. In a series of studies, Webb (1989) showed that although during mutual reasoning all participants benefit from the discourse, the one who delivers the explanations benefits even more than the one who listens to it. Elaborated explanations based on detailed clarifications and multiple sources of information or representations had the strongest effect on mathematics achievement. Furthermore, King (1998) indicated that the level of questions asked during peer interactions influences the responder's cognitive level, so that thought-provoking questions elicit reflective thinking and other types of higher-order cognitive response.

Mathematical communication can also assist in discovering errors and misconceptions that otherwise would remain implicit. Sometimes students are making two mistakes that cancel each other out and thereby attain the correct answer; only through mathematical communication can those errors become overt. In other cases, students may develop on their own simple mathematical rules that often lead to mathematical misconceptions – e.g. "multiplication always makes things bigger". These misconceptions persist despite subsequent evidence and instruction to the contrary (Steen, 1999). Communication, either orally or in writing, can help teachers and other participants detect such mistakes and make mathematics friendlier to the user (Maher and Martino, 1997).

The importance of communication in mathematics education is not limited only to older student. The US National Council of Teachers of Mathematics (NCTM) clarifies that mathematics communication has to start at an early age, from kindergarten through to the end of high school and college. According to the NCTM communication standards (2000, p. 59), students should be able to:

- organise and consolidate their mathematical thinking through communication

- communicate their mathematical thinking coherently and clearly to peers, teachers, and others

- analyse and evaluate the mathematical thinking and strategies of others

- use the language of mathematics to express mathematical ideas precisely.

For example, Prytula (2012) claims that when students work in small groups, all claims need to have a reason, be explained and discussed; everyone has to have a chance to talk, justify, and prove his/her conception. Prytula concludes that pre- and in-service professional development needs to shift from mastery of skills to metacognition.

The NCTM is not the only institute that emphasises the importance of communication in mathematics classrooms. For example, the PISA framework states again and again the importance of students justifying their mathematical reasoning (OECD, 2004, 2013, 2014). According to the OECD, communication competencies are required at all levels, up to the highest level in which "students can formulate and precisely communicate their actions and reflections regarding their findings, interpretations, arguments, and the appropriateness of these to the original situations" (OECD, 2004, p. 55).

The importance of communication for CUN problem solving makes it plausible that mathematics communication also enhances some aspects of social skills which are very important in real life. According to the OECD (2004), social skills include, among other competencies, the ability to: 1) create, maintain and manage personal relationships with others; 2) co-operate in teamwork, share responsibilities, leadership and support others; and 3) manage and resolve problems or conflicts that arise in the group due to divergent needs, interests, goals or values. Communication is an important component of social skills because interacting with others requires the ability to present ideas coherently and listen to others, give and receive constructive feedback, understand the dynamics of debates, and be able to negotiate and sometimes also to give up. Resolving conflicts requires consideration of one's own and others' interests and needs and the generation of solutions in which both sides gain. Thus, a by-product of enhancing mathematics reasoning and communication might be the promotion of social skills, an important outcome in itself.

Conclusion

Complex, unfamiliar and non-routine (CUN) tasks must form the core of a mathematics education appropriate for innovative societies and students should be involved in solving such tasks as well as routine mathematical problems. Mathematical reasoning, creativity and communication are essential components for solving CUN problems. Developing these competencies should not be limited to gifted students or to high-level grades. On the contrary, they can be applied in all age groups and should be the cornerstone of any effective learning environments. How to design such environments is the topic of the next chapter.

References

Ash, D. (2004), "Reflective scientific sense-making dialogue in two languages: The science in the dialogue and the dialogue in the science", *Science Education*, Vol. 88(6), pp. 855-884.

Ben-Peretz, M. (1990), *The Teacher-Curriculum Encounter: Freeing Teachers from the Tyranny of Text*, The State University of New York Press, Albany.

Brooks, J.G. and M.G. Brooks (1993), *In Search of Understanding: The Case for Constructivist Classrooms*, Association for Supervision and Curriculum Development, Alexandria, VA.

Csikszentmihalyi, M. (1996), *Creativity : Flow and the Psychology of Discovery and Invention*, Harper Perennial, New York.

Guilford, J.R. (1967), *The Nature of Human Intelligence*, McGraw Hill, New York.

Iluz, S., T. Michalsky and B. Kramarski (2012), "Developing and assessing the Life Challenges Teacher Inventory for teachers' professional growth", *Studies in Educational Evaluation*, Vol. 38(2), pp. 44-54.

King, A. (1998), "Transactive peer tutoring: Distributing cognition and metacognition", *Educational Psychology Review*, Vol. 10(1), pp. 57-74.

Leikin, R. and D. Pitta-Pantazi (2013), "Creativity and mathematics education: The state of the art ", *ZDM International Journal on Mathematics Education*, Vol. 45(2), pp. 159-166.

Light, P.H. and Z.R. Mevarech (1992), "Cooperative learning with computers: An introduction", *Learning and Instruction*, Vol. 2(3), pp. 155-159.

Maher, C.A. and A.M. Martino (1997), "Conditions for Conceptual Change: From Pattern Recognition to Theory Posing", in H. Mansfield and N. H. Pateman (eds.), *Young Children and Mathematics: Concepts and their Representation*, Australian Association of Mathematics Teachers, Sydney, Australia.

NCTM (National Council of Teachers of Mathematics) (2000), *Principles and Standards for School Mathematics*, NCTM, Reston, VA.

NCTM (1991), *Professional Standards for School Mathematics*, NCTM, Reston, VA.

New Jersey Mathematics Coalition and the New Jersey Department of Education (1996), *New Jersey Mathematics Curriculum Framework, The first four standards, standard 4- reasoning, K-12 overview*, State of New Jersey Department of Education.

OECD (2014), *PISA 2012 Results: Creative Problem Solving: Students' Skills in Tackling Real-Life Problems* (Volume V), PISA, OECD Publishing. http://dx.doi.org/10.1787/9789264208070-en.

OECD (2013), *PISA 2012 Assessment and Analytical Framework: Mathematics, Reading, Science, Problem Solving and Financial Literacy*, OECD Publishing. http://dx.doi.org/10.1787/9789264190511-en.

OECD (2004), *The PISA 2003 Assessment Framework: Mathematics, Reading, Science and Problem Solving Knowledge and Skills*, Education and Skills, PISA, OECD Publishing, Paris, *http://dx.doi.org/10.1787/9789264101739-en*.

Prytula, M.P. (2012), "Teachers' metacognition within the professional learning community", *International Education Studies*, Vol. 5(4).

Resnick, L.B. (1987), *Education and Learning to Think*, National Academy Press, Washington, DC.

Schoenfeld, A.H. (1992), "Learning to think mathematically: Problem solving, metacognition, and sense-making in mathematics", in D.A. Grouws, (ed.), *Handbook for Research on Mathematics Teaching*, MacMillan Publishing, New York, pp. 334-370.

Sheffield, L.J. (2013), "Creativity and school mathematics: Some modest observations", *ZDM International Journal on Mathematics Education*, Vol. 45(2), pp. 325-332.

Silver, E.A. (1997), "Fostering creativity through instruction rich in mathematical problem solving and problem posing", *ZDM International Journal on Mathematics Education*, Vol. 29(3), pp. 75-80.

Steen, G.J. (1999), "Genres of discourse and the definition of literature", *Discourse Processes*, Vol. 28(2), pp. 109-120.

Sternberg, R.J. and J.E. Davidson (eds.) (1995), *The Nature of Insight*, MIT Press, London.

Torrance, E.P. (1966), *The Torrance Tests of Creative Thinking Norms-Technical Manual Research Edition: Verbal Tests, Forms A and B; Figural Tests, Forms A and B*, Personnel Press, Princeton, NJ.

Wallas, G. (1926), *The Art of Thought*, Harcourt, Brace, New York.

Yan, Z. and F. Lianghuo (2006), "Focus on the representation of problem types in intended curriculum: A comparison of selected mathematics textbooks from mainland China and the United States ", *International Journal of Science and Mathematics Education*, Vol. 4(4), pp. 609-629.

Webb, N.M. (1989), "Peer interaction and learning in small groups", *International Journal of Educational Research*, Vol. 13(1), pp. 21-39.

Chapter 2

What is metacognition?

The term metacognition was first introduced to indicate the process of "thinking about thinking". Since then the concept has been elaborated and refined, although the main definition has broadly remained the same. Metacognition is now recognised to have two main components: "knowledge of cognition" (declarative, procedural and conditional knowledge), and the more important "regulation of cognition" (planning, monitoring, control and reflection). Basic metacognitive skills appear to start to develop in very young children and grow in sophistication with age and intellectual development. It is not yet clear how far metacognitive abilities in one domain can be transferred into another, but there is a strong relationship between metacognition and schooling outcomes with implications for educators, researchers and policy makers.

Recognising the types of mathematics problems appropriate for innovation-driven societies, and identifying the skills and cognitive processes suitable for solving them, only provide part of the picture. Performing complex tasks requires a higher-order "program" that acts as a cognitive engine to start the process, regulate the cognitive functioning and evaluate the product and the whole course of action. This cognitive engine receives information from the object level, processes it, debugs errors (when errors are identified), evaluates the execution and provides information back to the object level for further elaboration (Nelson and Narens, 1990). Flavell (1979) named these processes "metacognition" to emphasise their "meta" properties, where "meta" is used to mean *about* or *beyond* or *higher* than its own category. Hence, metacognition means "thinking about thinking" (Flavell, 1979) or a "person's cognition about cognition" (Wellman, 1985, p. 1).

Hence, metacognition is a form of cognition, a second or higher-order thinking process which involves active control over cognitive processes. It enables learners to plan and allocate learning resources, monitor their current knowledge and skill levels, and evaluate their learning level at various points during problem solving, knowledge acquisition or while achieving personal goals.

Flavell (1976) provides some useful examples explaining the concept of metacognition:

I am engaging in metacognition if I notice that I am having more trouble learning A than B; if it strikes me that I should double-check C before accepting it as a fact; (…) if I become aware that I am not sure what the experimenter really wants me to do; if I think to ask someone about E to see if I have it right. (Flavell, 1976, p. 232).

What is the difference between cognition and metacognition?

From the first introduction of the concept of metacognition, researchers have pointed out that although there is a large overlap between cognition and metacognition, the concepts are still different (e.g. Flavell, 1979; Brown, 1987, p. 66). For example, recalling your credit card's pin number is cognitive, but being aware of the strategy that would assist you to recall it is considered metacognitive. Solving an equation is a cognitive function, whereas reflecting on the answer and realising that the solution obtained does or does not fit the givens in the problem, is part of the metacognitive processes.

These examples may seem clear, but the distinction is sometimes more elusive. Because of the interchangeability of cognitive and metacognitive functions, a particular activity can be seen as either cognitive or metacognitive. Flavell (1979) assumes that metacognition and cognition differ in their content and functions, but are similar in their form and quality, i.e. both can be acquired, be forgotten, be correct or incorrect, be subjective, be shared, or be validated. Yet, while the content of cognition is the problem itself and the function is the solution execution, the contents of metacognition are the thoughts and its function is regulation of the thoughts (Hacker, 1998; Vos, 2001).

The relationships between cognition and metacognition have been studied by Veenman and his colleagues (e.g. Veenman et al., 1997; Veenman and Beishuizen, 2004; Veenman and Spaans, 2005; Veenman, 2013). In a series of studies involving students at various age levels, Veenman reported medium to high correlations between cognition and metacognition. Also in the area of mathematics, Van der Stel, Veenman, Deelen, and Haenen (2010) showed that in upper primary and early secondary school students, metacognition and intellectual ability are moderately correlated. Moreover, these researchers reported that in both age groups, metacognition has its own unique contribution to mathematics performance, on top of intellectual ability.

Models of metacognition

Research in the area of metacognition has flourished in recent decades (e.g. Stillman and Mevarech, 2010). It has included the development of theoretical models (e.g. Flavell, 1979; Brown, 1987; Nelson and Narens, 1990; Schraw at el., 2006; Veenman, 2013), empirical and quasi-experimental studies and intervention programmes.

Defining metacognition led researchers to construct models that clarify the specific components of metacognition and the relationships among them. Below is a short review of the major models of metacognition.

Flavell's model of cognitive monitoring

In his classic article "Metacognition and cognitive monitoring", Flavell (1979) makes the first attempt to define the components of metacognition by proposing a formal model of cognitive monitoring/regulation. His proposal includes four components: 1) metacognitive knowledge; 2) metacognitive experiences; 3) goals or tasks; and 4) actions or strategies (Figure 2.1).

Figure 2.1. **Flavell's model of metacognitive monitoring**

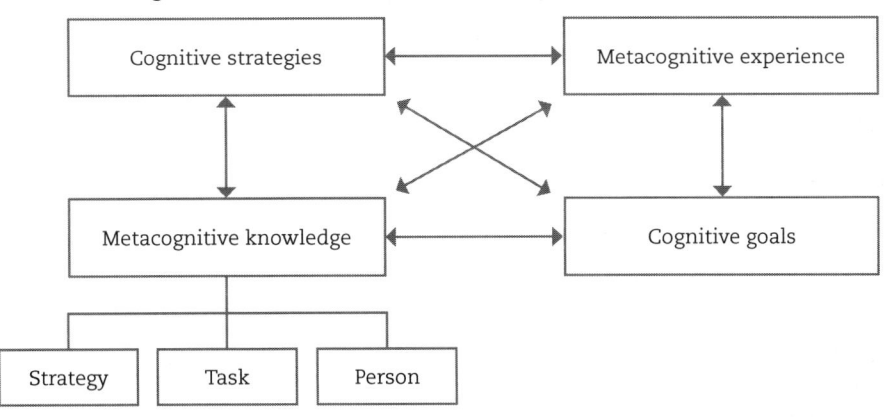

Source: Flavell, J. H. (1979), "Metacognition and cognitive monitoring: A new area of cognitive-developmental inquiry", *American Psychologist*, Vol. 34(10), pp. 906-911.

Metacognitive knowledge is defined by Flavell as one's knowledge or beliefs about the factors that relate to cognitive activities. The distinction between cognitive and metacognitive knowledge is not very clear and usually lies in the use of the knowledge or the object of the knowledge, rather than in the kind of the knowledge itself. Metacognitive knowledge is perceived as leading the individual to be engaged in or abandon the task, and thus it usually perceives or follows the cognitive activities. For example, judging the given task as difficult and one's relevant competencies as poor may result in one giving up the task, or in contrast in investing more effort in performing it. Because of that, Flavell (1979) assumes that metacognitive knowledge is the main category regulating cognitive performance.

Flavell (1979) identifies three categories of metacognitive knowledge: *person, task* and *strategy*. The *person* category comprises all the knowledge and beliefs that one has about oneself and others as cognitive processors. In the same vein, the *task* category refers to one's knowledge and beliefs about the nature of the given task and its demands: is it difficult or easy? Does it include all the information necessary for solving it? Or, is the question (or demand) clearly stated? Finally, the *strategy* category includes the identification of the task's goals and the knowledge about which cognitive processes are likely to be effective for solving the task. According to Flavell, although these three categories are independent, they nevertheless work together when one attempts a solution. The following case exemplifies how the metacognitive knowledge regulates cognitive performance:

> Ruth was asked to remember a telephone number. She knew that it is important to remember the seven digits in the correct order (*task* category), but she believes that her memory competencies are poor (*person* category), and therefore she has to look for strategies that would assist her in recalling the number (*strategy* category). The actual recall of the number represents the cognitive process (Flavell, 1979).

This example makes it clear that when a person is aware of the task's demands and his or her personal competencies, even a rote learning task such as the one described above relies on the activation of metacognitive processes. While rote learning is quite often executed automatically or unconsciously without going through the metacognitive path, this is not the case when performing complex tasks, such as CUN problems. The very definition of complex tasks means that the solution cannot be obtained by the automatic application of ready-made algorithms. In complex tasks, the solver has to estimate the task difficulty level with respect to his or her competencies, and decide what to do accordingly. Hence, in solving complex tasks, the activation of metacognitive processes is unavoidable.

Metacognitive experience, the second major category, refers to the conscious or unconscious processes that accompany any success or failures in learning or performing a cognitive enterprise, for example a feeling of confusion after reading a passage, or a feeling of success after solving a complicated maths task. Such experiences occur at any stage of the task's performance, and thus may influence the present or future cognitive performance. The metacognitive experiences can lead the individual to invest more time and mental energy in the task, or in contrast be frustrated and abandon the task. Flavell (1979) and later Efkelides (2011) conclude that these experiences are more likely to happen in situations that demand careful and highly conscious, reflective thinking (for more information on metacognitive experiences, see Chapter 6).

Cognitive goals, the third major category, refer to the actual objectives of a cognitive endeavor, such as reading and understanding a passage for an upcoming quiz, or being able to solve a multiple-stage word problem. The cognitive goals trigger the use of metacognitive knowledge, which in turn activates the other metacognitive components.

Finally, the *strategies or actions* refer to the use of specific techniques that may assist in achieving these goals, such as remembering that presenting the data in a table had helped increase comprehension in the past.

The four major categories in Flavell's model influence one another either directly or indirectly, and thus monitor and control the cognitive functions (Flavell, 1979). Yet, we all know that knowledge of cognition does not guarantee regulation of cognition. The very fact that one knows how the brain works or how to monitor cognitive processes, does not necessarily result in actual monitoring and control. Thus, a decade after Flavell introduced his metacognitive model, Brown (1987) proposed a new model which distinguished between the knowledge of cognition and its regulation.

Brown's model of metacognitive knowledge and regulation

Brown (1987) divides metacognition into two broad categories: 1) knowledge of cognition, and 2) regulation of cognition. Knowledge of cognition is defined as activities that involve conscious reflection on one's cognitive abilities and activities. Regulation of cognition refers to self-regulatory mechanisms used during an ongoing attempt to learn or solve problems. According to Brown, although these two forms of metacognition are closely related, each feeding the other recursively, they can be readily distinguishable.

Knowledge of cognition denotes the information people have about their own cognitive processes. It is based on the assumption that learners can step back and consider their own cognitive processes as objects of thought and reflection. Learners know, for example, that they have to reread a text in order to recall it, or that underlining the important ideas during reading is an effective strategy, or that drawing a graph assists in identifying trends.

Regulation of cognition consists of the activities used to regulate and oversee learning. These processes include *planning* activities prior to undertaking learning (e.g. predicting outcomes and scheduling strategies); *monitoring* activities during learning (monitoring, testing, revising and rescheduling one's strategies for learning); and checking solutions by *evaluating* the outcome of any strategic actions against criteria of efficiency and effectiveness. This model emphasises the executive processes, stressing the importance of the control that people bring or fail to bring to cognitive endeavors.

Schraw's model of metacognition

In the mid-1990s and the beginning of the 2000s, Schraw and Dennison (1994) proposed a model that elaborates on Brown's model of metacognition. It uses the same two basic components: knowledge of cognition and regulation of cognition, broken down into various subcategories. *Knowledge of cognition* includes three components: 1) *declarative knowledge* (knowledge about ourselves as learners and what factors influence our performance); 2) *procedural knowledge* (knowledge about strategies and other procedures appropriate for solving the problem or enhancing learning); and 3) *conditional knowledge* (knowledge of why and when to use a particular strategy). *Regulation of cognition* includes the same three basic components proposed by Brown: planning, monitoring and evaluation.

In order to validate the model, Schraw and Dennison (1994) administered a 56-item questionnaire to university students. Factor analysis revealed two additional components: *information management* and *debugging errors*. Hence, in the model of Schraw and Dennison, planning involves goal setting, allocation of resources, choosing appropriate strategies and budgeting time. *Monitoring* includes self-testing skills necessary to control learning, and debugging errors when they are diagnosed. *Information management* includes one's ability to organise, classify and retrieve information. *Evaluation* refers to assessing the products, reviewing the learning processes, and re-evaluating one's goals. Usually, planning takes place prior to learning, whereas monitoring, control, debugging and information management are activated during learning, and evaluation immediately after learning.

The distinction between metacognitive knowledge and regulation has led to important changes in how we currently perceive metacognition. Until recently, most of the studies in the area of metacognition made the assumption that metacognition has to be conscious and verbal. Consequently, researchers assumed that knowledge and self-monitoring need to be articulated in order to be considered as metacognition. However, recent studies have started to ask whether regulation has to always be conscious. If a child cannot describe her thoughts, does it mean that she did not monitor and control her problem solving processes? Parents, kindergarten teachers, and rigorous studies based on observations (e.g. Whitebread, 1999) have suggested that even young children, who cannot articulate their thinking, nevertheless plan ahead, monitor, control and evaluate their activities when the task fits the child's competencies and interests. Sangster-Jokic and Whitebread (2011) claimed that "some processes involved in metacognitive control might not always be available to consciousness or stored as articulated knowledge". This argument has prompted a shift in assumptions and a more inclusive conceptualisation of metacognition which argues that "both conscious and implicit forms of learning in relation to metacognitive processes need to be acknowledged in order to obtain a fuller understanding of metacognition and the manner in which it develops in children." (Sangster-Jokic and Whitebread, 2011, p.82).

Although the importance of metacognition is widely recognised (e.g. Veenman et al., 2006), confusion and ambiguity have arisen due to the different conceptualisations of metacognition, the use of a single term to describe different phenomena (e.g. Flavell versus Brown), and the unclear distinction between cognition and metacognition. Metacognition has become a wide umbrella incorporating a large number of different processes and skills.

General versus domain-specific metacognition

An issue of particular importance to educators and researchers is whether metacognition is general, or rather task and domain specific. Teachers often ask themselves if students who know how to monitor and control problem solving in mathematics would also be able to regulate their reading comprehension. Similarly, researchers raise the question of the extent to which facilitating regulation in one specific domain would be transferred to another domain. These issues have practical implications because if metacognition is general, it could be taught in one learning situation and students might be expected to transfer it to new situations, whereas domain-specific metacognition would have to be taught for each task or domain separately.

Findings regarding this issue are inconsistent. While several studies indicate that monitoring skills are general by nature (e.g. Schraw and Nietfeld, 1998), others provide evidence supporting the domain-specific approach (e.g. Kelemen, Frost and Weaver, 2000), and still others report strong relationships between general and domain-specific metacognitive knowledge (e.g. Neuenhaus, Artelt, Lingel and Schneider, 2010). It is possible, as Brown (1987) suggested, that knowledge about cognition is more general, relatively consistent within individuals and develops later, whereas regulation of cognition is more context dependent, changes from situation to situation, is affected by variables such as motivation and self-concept, and generally less accessible to conscious processes. Recent studies have shown, however, that under effective learning environments, students are able to transfer their metacognitive skills from one context to another (Mevarech and Amrany, 2008), or from one domain to another (Mevarech, Michalsky and Sasson, submitted).

How does metacognition develop with age?

In general, researchers disagree over the earliest age at which metacognition can be activated. While the early studies of metacognition argue that students can perform metacognitive activities only towards the end of elementary school, others claim that metacognitive knowledge and skills already develop during preschool or early-school years (for an excellent review see Veenman et al., 2006). There may be several reasons for these contradictory findings.

First, metacognitive ability is not something that one either possesses or does not possess, but rather extends along a continuum. It is quite possible that kindergarten children acquire metacognition at a very basic level, but that it becomes more sophisticated and academically oriented throughout life. Mevarech (1995) reported how kindergarten children (4 to 5-year-olds) activated metacognitive knowledge during maths problem solving. Whitebread (1999) used natural observations to describe preschoolers' (3 to 5-year-olds) metacognition, and Shamir, Mevarech and Gida (2009) report the way kindergarten children describe to their peers which strategies to apply in recalling tasks. Veenman et al. (2006), on the other hand, show that metacognition emerges at the age of eight to ten, and expands during the years thereafter (Berk, 2003; Veenman and Spaans, 2005; Veenman et al., 2006). In a series of studies these researchers provided evidence that metacognitive knowledge develops along a monotonic incremental line throughout the school years, parallel to the development of students' intellectual abilities. Schraw et al. (2006) indicate that most adults have metacognitive knowledge and can plan accordingly.

Second, metacognition includes multiple components. Thus, assessing one component may not reflect the abilities regarding another component (Berk, 2003; Veenman and Spaans, 2005; Veenman et al., 2006). Brown (1987), for example, argues that knowledge of cognition develops at a later age than regulation of cognition. It seems that certain metacognitive skills, such as monitoring and evaluation appear to mature later than others, such as planning, probably because children at school are less exposed to these processes (e.g. Focant, Gregoire and Desoete, 2006). Roebers et al. (2009) show that 9-year-old children revealed well-developed monitoring skills, but were less able to control their problem-solving processes than 11-12 year-old students. This conclusion is supported by recent studies showing that during mathematics problem solving, 8-year-old children exhibited greater planning and evaluation than self-monitoring (Kramarski, Weisse and Koloshi-Minsker, 2010).

Finally, widening the definition of metacognition to include nonverbal and unconscious activities enables researchers to document metacognition in very young children (e.g. 3-5 year-olds) when the tasks fit the child's capabilities and interests (Whitebread and Coltman, 2010). While metacognitive awareness is evident at the age of 4 to 6 years as a feeling that something is wrong (Blöte, Van Otterloo, Stevenson and Veenman, 2004; Demetriou and Efklides, 1990), it has been repeatedly shown that preschool and kindergarten children overestimate their own performance across a wide range of contexts (Schneider, 1998). In two experiments carried out with 4 to 6-year olds, Schneider (1998) showed that children's over-predictions were due to wishful thinking rather than to poor metacognition.

In summary, while the first studies on metacognition assumed that children cannot apply metacognitive processes in problem solving, the more recent studies show that 1) metacognition emerges at a very early age (around 3 years old); and 2) metacognition develops as a function of children's age. When a task fits the child's interest and capabilities, even preschool children can plan ahead, monitor their activities, and reflect on the processes and the outcomes. However, many questions are still open. How can metacognition be assessed in young children who cannot articulate their thinking? What tasks are appropriate for the younger age groups? What are the conditions that encourage young children to activate metacognitive processes?

How does metacognition affect learning and achievement?

Studies highlight the relationship between metacognition and academic achievement. In the last decade, it has become widely accepted that metacognition plays a crucial role in school achievement and beyond (Boekaerts and Cascallar, 2006; Sangers-Jokic and Whitebread, 2011). Children and young people with higher levels of metacognitive skills are more likely to succeed academically than students showing low levels of metacognition (Duncan et al., 2007; McClelland et al., 2000). Veenman et al. (2006) demonstrated that metacognition predicts school achievement in various academic areas and in different grade levels, even when intellectual ability is controlled. Reviewing what influences learning, Veenman et al. (2006) cited a review study by Wang, Haertel, and Walberg (1990) which "revealed metacognition to be a most powerful predictor of learning" (p. 3).

Similar findings were also reported in the area of mathematics. Stillman and Mevarech (2010a, 2010b) and Desoete and Veenman (2006) described a large number of studies that focused on metacognition and mathematics learning. In particular, metacognition relates to the solution of complex, unfamiliar and non-routine (CUN) problems more than to the solution of familiar, routine problems, probably because the later can be executed automatically by applying ready-made algorithms, whereas the solution of CUN problems requires the activation of the various components of metacognition (Mevarech et al., 2010). These findings apply not only to CUN tasks in mathematics, but to CUN tasks in other domains as well. For example, when reading a "simple" sentence one can comprehend it without consciously applying metacognitive strategies, whereas in reading a complicated text, the application of metacognitive strategies is essential (Carlisle and Rice, 2002).

Intensive research has also demonstrated that lower achievers and students with learning disabilities have deficits in monitoring and controlling their learning (e.g. Desoete, 2007). These students are likely to have difficulties in assessing their learning and in using metacognitive knowledge in solving the given problems (e.g. Efklides et al., 1999). Having negative metacognitive experiences may lead lower achievers to abandon tasks without even trying to attempt them (Paris and Newman, 1990).

An interesting study conducted by Sangster-Jokic and Whitebread (2011) showed the role of metacognition in assisting the monitor and control of motor performance of children with developmental co-ordination disorder (DCD). Reviewing the literature on developmental co-ordination disorder, the authors concluded that "the examination of self-regulation and metacognitive competence is a promising area for further understanding the difficulties of children with DCD" (p. 93).

Conclusion

The concept of metacognition has caused a change in the way we understand learning, mainly by shifting attention from the cognitive to metacognitive processes, and from the application of algorithms to "thinking about thinking", particularly to the importance of the planning , monitoring, control and reflective systems that regulate one's cognitive activities. The implementation of the "metacognitive engine" is essential in performing cognitive tasks including those in the area of mathematics.

Principles emerging from the studies of metacognition have significant implications for education. The main conclusions are:

- Although cognition and metacognition are different entities, they are closely related. Students who apply metacognitive processes tend to be higher achievers and vice versa; higher achievers usually apply metacognitive processes in learning and problem solving.

- According to Flavell, metacognition refers to the task, person and strategies. In teaching problem solving, teachers and students have to bring up these three elements and the relationships between them.

- Regulation of cognition includes: planning, monitoring, control, reflection, error debugging (evaluation) and information processing. Teachers should consider incorporating all of these competencies in their ongoing teaching.

- The importance of regulating cognition means students need to develop the active implementation of metacognitive skills, rather than having only theoretical knowledge about them.

- Given the importance of domain-specific metacognition, learning environments have to provide students with domain-specific metacognitive tools.

- Most recent studies indicate that even young children can apply metacognitive processes when the tasks fit their interest and capabilities. This is an important finding showing that metacognitive processes can be applied at all age levels, and in various types of tasks (routine and CUN).

References

Berk, L.E., (2003), *Child Development*, 6th Edition, Allyn and Bacon, Boston.

Boekaerts, M. and E. Cascallar (2006), "How far have we moved toward the integration of theory and practice in self-regulation?" *Educational Psychology Review*, Vol. 18(3), pp. 199-210.

Blöte, A.W., S. G. Van Otterloo, C. E. Stevenson and M.V.J. Veenman (2004), "Discovery and maintenance of the many-to-one counting strategy in 4-year olds: A microgenetic study", *British Journal of Developmental Psychology*, Vol. 22(1), pp. 83-102.

Brown, A. (1987), "Metacognition, executive control, self-regulation and other more mysterious mechanisms", in F. Weinert and R. Kluwe (eds.), *Metacognition, Motivation and Understanding*, Erlbaum, Hillsdale, NJ, pp. 65-116.

Carlisle, J. F. and M.S. Rice (2002), *Improving Reading Comprehension: Research-Based Principles and Practice*, York Press, Timoniun, MD.

Desoete, A. (2007), "Evaluating and improving the mathematics teaching-learning process through metacognition", *Electronic Journal of Research in Educational Psychology*, Vol. 5(3), pp. 705-730.

Desoete, A. and M.V.J. Veenman (2006), "Introduction", in A. Desoete and M.V.J. Veenman (eds.), *Metacognition in Mathematics Education*, Nova Science Publishers, New York, pp. 1-10.

Demetriou, A. and A. Efklides (1990), "The objective and subjective structure of metacognitive abilities from early adolescence to middle age", in H. Mandl, E. De Corte, N. Bennett and H.F. Friedrich (eds.), *Learning and Instruction: European Research in an International Context*, Pergamon, Oxford, pp. 161-180.

Duncan, G.J. et al. (2007), "School readiness and later achievement", *Developmental Psychology*, Vol. 43(6), pp. 1428-1446.

Efklides, A. (2011), "Interactions of metacognition with motivation and affect in self-regulated learning: The MASRL model", *Educational Psychologist*, Vol. 46(1), pp. 6-25.

Efklides, A., A. Samara and M. Petropoulou (1999), "Feeling of difficulty: An aspect of monitoring that influences control", *European Journal of Psychology of Education*, Vol. 14(4), pp. 461-476.

Flavell, J.H. (1979), "Metacognition and cognitive monitoring: A new area of cognitive-developmental inquiry", *American Psychologist*, Vol. 34(10), pp. 906-911.

Flavell, J.H. (1976), "Metacognitive aspects of problem solving", in L.B. Resnick (ed.), *The Nature of Intelligence*, Erlbaum, Hillsdale, NJ, pp. 231-236.

Focant, J., J. Grégoire and A. Desoete (2006), "Goal-setting, planning and control strategies and arithmetical problem solving at grade 5", in M.J. Veenman and

A. Desoete (eds.), *Metacognition in Mathematics Education,* Nova Sciences Publishers, New York, pp. 51-71.

Hacker, D.J. (1998), "Definitions and empirical foundations", in D.J. Hacker, J. Dunlosky and A.C. Graesser (eds.), *Metacognition in Educational Theory and Practice,* Lawrence Erlbaum Associates, pp. 1-23.

Kelemen, W.L., P.J. Frost and C.A. Weaver III (2000), "Individual differences in metacognition: Evidence against a general metacognitive ability", *Memory and Cognition,* Vol. 28(1), pp. 92-107.

Kramarski, B., I. Weiss and I. Kololshi-Minsker (2010), "How can self-regulated learning support the problem solving of third-grade students with mathematics anxiety?" *ZDM International Journal on Mathematics Education,* Vol. 42(2), pp. 179-193.

McClelland, M.M., F.J. Morrison and D.L. Holmes (2000), "Children at risk for early academic problems: The role of learning-related social skills", *Early Childhood Research Quarterly,* Vol. 15(3), pp. 307-329.

Mevarech, Z.R. (1995) "Metacognition, general ability and mathematical understanding in young children", *Early Education and Development,* Vol. 6(2), pp. 155-168.

Mevarech, Z.R. and C. Amrany (2008), "Immediate and delayed effects of metacognitive instruction on regulation of cognition and mathematics achievement", *Metacognition and Learning,* Vol 3(2), pp. 147-157.

Mevarech, Z.R., S. Terkieltaub, T. Vinberger and V. Nevet (2010), "The effects of meta-cognitive instruction on third and sixth graders solving word problems", *ZDM International Journal on Mathematics Education,* Vol. 42(2), pp. 195-203.

Mevarech, Z. R., T. Michalsky, H. Sasson (submitted), "The effects of different self-regulated learning (SRL) interventions on students' science competencies", Metacognition Special Interest Group biannual meeting, Istanbul.

Nelson, T.O. and L. Narens (1990), "Metamemory: A theoretical framework and new findings", in G.H. Bower (ed.), *The Psychology of Learning and Motivation,* Vol. 26, Academic Press, New York, pp. 1-45.

Neuenhaus, N. et al. (2011), "Fifth graders metacognitive knowledge: General or domain-specific?" *European Journal of Psychology of Education,* Vol. 26(2), pp. 163-178, DOI 10.1007/s10212-010-0040-7.

Paris, S.G. and R.S. Newman (1990), "Developmental aspects of self-regulated learning", *Educational Psychologist,* Vol. 25(1), pp. 87-102.

Roebers, C.M., C. Schmid and T. Roderer (2009), "Metacognitive monitoring and control processes involved in primary school children's test performance", *British Journal of Educational Psychology,* Vol. 79(4), pp. 749-767.

Sangster Jokic, C. and D. Whitebread (2011), "The role of self-regulatory and metacognitive competence in the motor performance difficulties of children with developmental coordination disorder: A theoretical and empirical review", *Educational Psychology Review*, Vol. 23, pp. 75-98.

Schneider, W. (1998), "Performance prediction in young children: Effects of skill, metacognition and wishful thinking", *Developmental Science*, Vol. 1(2), pp. 291-297.

Schraw, G. and R.S. Dennison (1994), "Assessing meta-cognitive awareness", *Contemporary Educational Psychology*, Vol. 19(4), pp. 460-475.

Schraw, G. and J. Nietfeld (1998), "A further test of the general monitoring skill hypothesis", *Journal of Educational Psychology*, Vol. 90(2), pp. 236-248.

Schraw, G., K.J. Crippen and K. Hartley (2006), "Promoting self-regulation in science education: metacognition as part of a broader perspective on learning", *Research in Science Education*, Vol. 36, pp. 111-139.

Shamir, A., Z.R. Mevarech and H. Gida (2009), "The assessment of young children's metacognition in different contexts: Individualised vs. peer assisted learning", *Metacognition and Learning*, Vol. 4(1), pp. 47-61.

Stillman, G. and Z.R. Mevarech (eds.) (2010a), "Metacognition research in mathematics education", *ZDM International Journal on Mathematics Education*, Special Issue, Vol. 42(2),

Stillman, G. and Z.R. Mevarech (2010b), "Metacognition research in mathematics education: From hot topic to mature field", *ZDM International Journal on Mathematics Education*, Vol. 42(2), pp. 145-148.

Van der Stel, M. and M.V.J. Veenman (2010), "Development of metacognitive skilfulness: A longitudinal study", *Learning and Individual Differences*, Vol. 20(3), pp. 220-224.

Veenman, M.V.J. (2013), "Metacognition and learning: Conceptual and methodological considerations revisited. What have we learned during the last decade?", keynote speech, 15th Biennial EARLI Conference for Research on Learning and Instruction, Munich, 27-31 August.

Veenman, M.V.J. and J.J. Beishuizen (2004), "Intellectual and metacognitive skills of novices while studying texts under conditions of text difficulty and time constraint", *Learning and Instruction*, Vol. 14(6), pp. 619-638.

Veenman, M.V.J., P. Wilhelm and J.J Beishuizen (2004), "The relation between intellectual and metacognitive skills from a developmental perspective", *Learning and Instruction*, Vol. 14(1), pp. 89-109.

Veenman, M.V.J. and M.A. Spaans (2005), "Relation between intellectual and metacognitive skills: Age and task differences", *Learning and Individual Differences*, Vol. 15(2), pp. 159-176.

Veenman, M.V.J., J.J. Elshout and J. Meijer (1997), "The generality vs. domain-specificity of metacognitive skills in novice learning across domains", *Learning and Instruction*, Vol. 7(2), pp. 197-209.

Veenman, M.V.J., B.H.A.M. Van Hout-Wolters and P. Afflerbach (2006), "Metacognition and learning: conceptual and methodological considerations", *Metacognition and Learning*, Vol. 1, pp. 3-14.

Vos, H. (2001), "Metacognition in higher education", PhD Thesis, University of Twente. http://doc.utwente.nl/37291/1/t0000011.pdf.

Wang, M.C., G.D. Haertel and H.J. Walberg (1990), "What influences learning? A content analysis of review literature", *Journal of Educational Research*, Vol. 84(1), pp. 30-43.

Wellman, H.M. (1985), "Origins of metacognition", in D.L.F. Pressley, G.E. McKinnon, and T.G. Waller (eds.), *Metacognition, Cognition and Human Performance,* Vol. 1, Academic Press, Orlando, Florida.

Whitebread, D. (1999), "Interactions between children's metacognitive abilities, working memory capacity, strategies and performance during problem-solving", *European Journal of Psychology of Education*, Vol. 14(4), pp. 489-507.

Whitebread, D. and P. Coltman (2010), "Aspects of pedagogy supporting metacognition and self-regulation in mathematical learning of young children: Evidence from an observational study", *ZDM International Journal on Mathematics Education*, Vol. 42(2), pp. 163-178.

Chapter 3

Metacognitive pedagogies

Can metacognition be taught? And if so, what are the conditions that can facilitate metacognitive application in the classroom? While the research shows that metacognition can be successfully taught, implicit guidance is not enough. Co-operative learning should help to foster metacognition by providing ample opportunities for students to articulate their thinking and be involved in mutual reasoning, nevertheless students still have to be taught how to apply these processes and also intensively practise them. Effective metacognitive guidance needs to be explicit, embedded in the subject matter, involve prolonged training, and inform learners of its benefits. A number of methodologies for teaching metacognition have been developed, all of which use social interactions and self-directed questioning in order to encourage learners to be aware of their metacognitive processes and apply these processes in learning.

Research has shown high positive correlations between problem solving and metacognition in various areas, including mathematics (e.g. De Corte et al., 2000; Desoete and Veenman, 2006; Kramarski and Zoldan, 2008; Stillman and Mevarech, 2010), reading (e.g. Palincsar and Brown, 1984), science (e.g. Zion et al., 2005) and even physical co-ordination (Kitsantas and Kavussanu, 2011). The positive relationship between metacognition and school achievement raises the question of "what causes what": does metacognition facilitate improved schooling outcomes or vice versa? If indeed metacognition affects schooling outcomes, there is reason to suppose that teaching metacognitive skills would enhance achievement. This raises further questions: what does metacognition look like in the classroom and what are the necessary conditions for applying metacognitive pedagogies? Are co-operative learning and the use of information and computer technologies (ICT) necessary or do they simply facilitate the teaching of metacognition? Do teachers need explicit training to promote metacognition or is implicit guidance sufficient? In short, how, when and for whom is metacognitive instruction needed?

Can metacognition be taught?

Whether metacognition can be taught is not at all self-evident; it reminds us of the debates about the extent to which IQ is teachable. Yet, in the case of metacognition, rigorous research has shown that teaching it is plausible not only in mathematics, but also in other domains, including reading, sciences and languages.

The research supports a few general conclusions. First, successful mathematics learners are metacognitively active (Schoenfeld, 1992). They think about what they are doing and why they are doing it, and they reflect on the learning outcomes. Second, it is possible to foster these metacognitive skills during the early years with positive benefits for academic achievement (Dignath and Buettner, 2008; Dignath, Buettner and Langfeldt, 2008; Fantuzzo et al., 2007). Finally, learning conditions and teachers have an important role in promoting cognitive and metacognitive processes (e.g. Cardelle-Elawar, 1995). The challenging issue is *how* metacognition can be taught, and what are its benefits and trade-offs.

What is the role of co-operative learning?

Many studies have emphasised the importance of a supportive social environment as an effective means to facilitate cognition and metacognition in learning (Lai, 2011; Lin, 2001). In particular, researchers have recommended the use of co-operative learning structures for fostering learning achievements (e.g. King, 1998; Slavin, 2010) and for enhancing metacognitive skills (Kuhn and Dean, 2004; Efklides, 2008; McLeod, 1997; Schraw and Moshman, 1995; Schraw et al., 2006). Steen (1999) claims that advocates of co-operative learning come from two different backgrounds: educators who evaluate these activities as effective for learning, and people outside the educational arena, in business, science, sport, music, etc., who view teamwork as essential for producing productive outcomes. However, many still raise questions of whether co-operative activities in themselves produce higher school achievements for the individual learner (e.g. Steen, 1999).

Co-operative learning consists of "small groups of learners working together as a team to solve a problem, complete a task, or accomplish a common goal" (Artzt and Newman, 1990, p.448). The term thus covers a number of teaching and learning methods, sometimes called peer-assisted learning (Fuchs et al., 2001), or team learning (Slavin, 2010). Learning in pairs is also often regarded as co-operative learning (Dansereau, 1988; King, 1998). A common characteristic of all co-operative learning methods is the division of the whole class into small learning groups of four to six students who have to complete a common task. Box 3.1 briefly describes the main co-operative learning methods used in mathematics classrooms.

Box 3.1. Co-operative learning methods used in mathematics classrooms

The main aim of all these methods is to increase each student's participation in the learning process and give them all equal opportunities for success (e.g. Slavin, 2010). To achieve these goals, classes are divided into pairs or small groups of three to six students. The differences relate to the role of the teacher, student activities, individual accountability and the evaluation process followed by reward for success.

Student Team-Achievement Divisions (STAD) (Slavin, 1994): In STAD, students are assigned to heterogeneous groups (or teams, in Slavin's terminology) of four or five members. The method uses a four-step cycle: 1) teach; 2) team study; 3) test and 4) recognition. First, the teacher presents the new concept(s) to the whole class, usually by using the lecture-discussion technique. Team members work co-operatively on the work sheets provided by the teacher, helping one another, and preparing themselves for the quiz that is taken individually. The teacher grades the quiz and compares the scores to the previous quizzes' scores of each individual team member. The team is rewarded according to the overall improvement of all team members. Each cycle takes three to five class periods. STAD has been successfully implemented in mathematics (and other subjects) classrooms from second grade to college.

Teams-Games-Tournament (TGT) (Slavin, 1994): TGT is similar to STAD, but instead of using weekly quizzes, the evaluation is based on weekly tournaments in which students compete against members of other teams with a similar past record in mathematics. The team score is based on the number of points each member brings to the team. As in STAD, teams are rewarded according to their improvement.

Team-Assisted Individualisation (TAI) (Slavin, 2010): TAI is especially designed for upper primary mathematics (grades 3 to 6), or older students who are not ready for algebra course. In contrast to STAD and TGT, in TAI each student is tested individually prior to the beginning of the study and paced according to his or her own abilities. The co-operative element comes from encouraging students to help one another with any problem. The team's weekly reward is based on the number of units each team member completed.

(continues...)

**Box 3.1. Co-operative learning methods used
in mathematics classrooms** (continued)

Peer-Assisted Learning (PAL) (Fuchs et al., 2001): in PAL, children learn in pairs. The pair members take turns as teacher and learner. Children are taught simple teaching strategies for helping each other. Pairs are rewarded according to both learners' scores on the quiz. PAL has been implemented successfully in primary and early secondary (middle school) mathematics classrooms.

Jigsaw (Aronson and Patnoe, 2011): in Jigsaw, each group member is responsible for learning and then teaching other group members a section of the unit to be studied. Aronson suggested using a five-step approach to implementing Jigsaw in the classroom: 1) assign students into small heterogeneous groups of three to six students; 2) divide the topic to be studied into sections or subtopics according to the number of students in the groups and allocating a section to each group member; 3) "expert groups" of students who were assigned to teach the same section work temporarily together to become experts on their section; 4) "expert" students return to their original groups and teach that section to the other group members; and 5) students are assessed on the whole topic. Groups are rewarded as in STAD. Jigsaw has been implemented at all education levels including tertiary education.

Learning Together (Johnson and Johnson, 1999): in this method, small heterogeneous groups of four to six students receive assignments that have to be solved together. The group hands in a single sheet and receives a team score based on the team performance. Before students start studying in small groups they are exposed to "team-building" activities that focus on intra-group discussions, giving constructive feedback, etc. Learning Together has been used in primary and secondary mathematics classrooms.

Group Investigation (Sharan and Sharan, 1992): in Group Investigation students choose their own group members of up to six people with whom they would like to work on an inquiry topic or a project. The topic/project is divided into subtopics on which team-members work. Each group then makes a presentation for the entire class.

Cooperative Mastery Learning (Mevarech, 1985, 1991): This method is similar to STAD, but following the weekly quizzes, students who did not master the topic receive remedial activities and the others are administered enrichment tasks. The remedial and enrichment activities are performed either co-operatively or individually with the teacher's assistance.

Source: Adapted from Slavin, (2010).

The different co-operative learning methods are rooted in different theoretical approaches. Piaget (1985) and Vygotsky (1978) highlight the potential of student interactions for enhancing cognitive development. According to Piaget, when a learner confronts contrasting facts or dissimilar phenomena he or she tends to resolve it in order to obtain equilibrium. Piaget coined this phenomenon "cognitive conflict". For example, in a classical experiment, children were asked to hypothesise if a piece of wood would sink or float in water. Most children said that it would sink. However, observing the wood float in the water, children were curious to resolve the conflict. Or, for example, in mathematics, many students mistakenly think that $(a+b)^2$ is equal to $2a+2b$. When they are asked to square $(2+3)$ they immediately realise that it is not equal to $2x2+2x3$, and try to find the source of the mistake by simplifying $(a+b)^2$ into $(a+b)x(a+b)$. In primary school, many children mistakenly think that multiplication "always makes bigger" and are surprised to see that the product of fractions smaller than one is smaller than each of the multipliers. The probability of cognitive conflicts arising is higher when students study together than when they study individually because each student brings his or her knowledge to the learning situation; that knowledge does not always coincide.

Vygotsky (1978) conceptualised learning (e.g. cognitive development) in a different way. He coined the term "zone of proximal development" as the distance between what an individual can attain alone and what he or she can accomplish with the help of a more capable other, either a peer or an adult. Group work provides ample opportunities for students to participate in mutual reasoning and conflict resolution. Cognitive conflicts may arise as students critically examine each other's reasoning and participate in group discussions. These in turn would encourage students to discuss the conflicts and suggest ways to resolve them (e.g. Artzt and Yaloz-Femia, 1999; McClain and Cobb, 2001; Mevarech and Light, 1992).

Yet, co-operative learning may have some drawbacks. For instance, group discussions may lead to a polarisation of positions instead of a productive exchange of ideas. A team member can convince all others to accept an erroneous concept as a correct one. Quite often, the group is too eager to start the solution without planning ahead, or the group is willing to finish the assignment without reflecting on the solution. Furthermore, lower achievers and shy students might not be involved in the group learning process. Sometimes in mixed gender groups the boys take over while the girls are left behind. It is only under certain conditions that these learning methods yield the desired outcomes (Slavin, 2010).

Kuhn and Dean (2004) argue that social discourse can cause students to "interiorise" processes by providing elaborations and explanations, which have been associated with improved problem solving outcomes. By justifying one's thinking and explaining it to others, and by challenging peers' explanations regarding the problem solution, learners can examine their own thinking and improve the efficacy of their own problem solving (King, 1998; Mevarech and Light, 1992; Mevarech and Kramarski, 1997; Webb, 2008). Studies that examine the behaviour of students in co-operative groups consistently find that students who give and receive comprehensive explanations are the ones who gain most from the co-operative setting both in terms of metacognitive skills and learning performance (King, 1998; Webb, 2008). In fact, these studies demonstrate that giving and receiving final answers without explanations is correlated negatively with achievement gains. Furthermore, these studies indicate that those who gave the explanations benefitted from the interaction even more than those who received them (Webb, 2008; Mevarech and Shabtay, 2012). The theory of metacognition clarifies this finding by suggesting that in order to give explanations one has to understand what the problem is all about, connect it to one's existing knowledge and the knowledge of other team members, suggest appropriate strategies, discuss various alternatives, and reflect on the solution process at all its stages (before, during and after solving the problem). Quite often, group discussions lead participants to think how to deliver the information so that it would be interpreted correctly by other group members (Hoppenbrouwers and Weigand, 2000) and how to eliminate the possibility of having to say "what I said is not what I meant" or "your interpretation of what I said is not what I meant". It seems, therefore, that group interactions may encourage students to provide explanations and use the language of mathematics correctly in articulating their reasoning (Kramarski and Dudai, 2009; Kramarski and Mizrachi, 2006; Mevarech and Kramarski, 1997).

In addition, several researchers argue that during the social interaction, a shared metacognitive experience emerges (Efklides, 2008; Lin, 2001) since group participants act as external regulators of their peers' cognitive, metacognitive and motivational behaviour. Hence, group discussions might enhance clarifications of students' understanding and can encourage the activation of metacognitive knowledge and cognitive regulation skills (Hadwin, Järvelä and Miller, 2011; Schraw and Moshman, 1995).

Indeed, co-operative learning has been well known and widely used worldwide for more than four decades (Slavin, 2010). In parallel with the intensive implementation of co-operative learning, researchers have examined its effects on various outcomes, including metacognition and mathematics achievements. This research has been summarised in several studies using meta-analytic techniques (Dignath and Buettner, 2008; Dignath et al., 2008; Hattie et al., 1996; Hattie, 1992; Marzano, 1998; Slavin, 2010). In particular, it is worth mentioning Slavin's studies based on "best-evidence syntheses" (e.g. Slavin and Lake, 2008; Slavin et al., 2009) in which Slavin and colleagues calculated the effect sizes of co-operative learning by selecting only studies that meet strict criteria. All these experimental studies showed co-operative learning methods to have overall positive effects on schooling outcomes, compared with control groups who studied individually.

However, research findings show that although co-operative learning is a natural setting for learners to supply explanations, elaborate their reasoning, and reflect on their own and others' solution processes, these processes have not always materialised spontaneously (e.g. King, 1998; Kramarski, Mevarech and Arami, 2002). For example, Steen (1999) indicates that while co-operative learning is effective for primary school children, for high school students and adults the evidence is more mixed. In contrast, Dignath and Buettner concluded from their meta-analysis that co-operative learning has positive effects on learning for middle and high school students, but has no effects or even negative effects for primary school pupils compared with "traditional" instruction (2008, p. 248). Dignath and Buettner explained this finding by considering students' experiences learning in small groups:

> It is obvious that the positive effects of cooperative learning can only surface if students know rules about how to behave when working in groups, it would not be enough to let students sit around a table in small groups without providing them with any systematic instruction. Hence, a possible reason for the negative effect of group work on training effects at primary school level might be that students were not used to working in groups and did not receive enough instruction about co-operative learning. Older students have

a higher probability of already knowing about co-operative working, since children develop co-operation skills during middle childhood (Cooper et. al., 1982) (Dignath and Buettner, 2008, pp. 256-257).

Slavin, one of the leading co-operative learning researchers admits that co-operative learning "has proven its effectiveness in hundreds of studies throughout the world ... yet observational studies (e.g. Antil et al., 1998) find that most use of cooperative learning is informal and does not incorporate the group goals and individual accountability that research has identified to be essential" (Slavin, 2010, p. 173). Slavin (2010) adds that co-operative learning has to be reshaped for the 21st century.

This situation raises the question of why learning in co-operative settings has not always fulfilled its potential. The main reason is that simply providing learners the opportunity to study together without guiding them how to monitor, control and evaluate their learning is not sufficient for promoting metacognition and mathematics problem solving. King (1998), Webb (2008), and many others who intensively studied students' behaviour in small groups, came to the conclusion that students' interactions are ineffective without the provision of metacognitive scaffolding. It seems, therefore, that for most students, implicit hints, such as "what are you doing here?" do not result in them applying metacognitive processes.

To summarise, while co-operative learning has the potential to facilitate learning by providing ample opportunities for students to articulate their thinking and be involved in mutual reasoning, by itself it is not sufficient to foster cognitive and metacognitive learning. Students do not spontaneously encourage one another to activate metacognitive processes in the small groups, nor do they always use the co-operative settings to advance their own learning and that of their peers. Hence, whether students study individually or in co-operative settings, they have to be explicitly taught how to apply metacognitive processes during learning.

Is explicit practice necessary?

We all know that "practice makes perfect". To be a champion in sport, chess, music, visual arts, science and other disciplines one has to practice in order to attain mastery. Interviewing big talents reveals the large number of hours they devote every day to practice (Bloom, 1985). Furthermore, most of the big talents practice with the aid of mentors or coaches who explicitly guide them in planning and regulating their behaviours. If champions need explicit metacognitive training and lots of practice, ordinary learners certainly need it as well.

Although a large body of evidence (Dignath and Buettner, 2008; Kistner, 2010) shows that implicit teaching of metacognition has only minor effects on students' behaviour and does not increase awareness, teachers tend to give students the freedom to self-regulate their learning. Actual observations in the classrooms have indicated that teachers do not explicitly train students in how to implement metacognitive processes and self-regulate their learning, probably because of teachers' beliefs that implicit training is appropriate (Dignath, 2012; Verschaffel et al., 2007). Findings indicate, however, that metacognitive training in the classroom has to be explicit and followed by intensive practice (Dignath and Buettner, 2008; Dignath et al., 2008).

Metacognitive pedagogies: how, when and for whom?

Given the benefits of explicit metacognitive training for primary and secondary school students (e.g. Dignath and Buettner, 2008), researchers and educators have started to design a variety of metacognitive pedagogies. These are methods to guide students to implement metacognition in learning, sometimes also called metacognitive guidance, metacognitive intervention, metacognitive instruction, or metacognitive scaffolding. These pedagogies are ecologically valid, i.e. the learning materials, methods and settings approximate real-life situations in the classrooms (e.g. Sangster Jokic and Whitebread, 2011).

Veenman et al. (2006) emphasise the key factors that are crucial in implementing metacognitive instruction:

1) Embedding metacognitive instruction in the subject content matter to ensure the connection of new knowledge to what students already know.

2) Informing learners about the usefulness of metacognitive activities to make them exert the initial extra effort.

3) Prolonged training to guarantee the smooth and maintained application of metacognitive activity.

4) Explicit guidance to ensure awareness and efficient implementation) added later, personal comment).

Although these metacognitive interventions vary considerably, each approach has three underlying components in common. First, the techniques rely on teachers' ability to train students to *be aware* and *consciously reflect* on their own thought processes while simultaneously emphasising the importance of mastering the material. Second, students gain metacognitive knowledge through social interactions with the classroom teacher and peers. Finally, self-directed questioning is effective for promoting metacognition because it guides the learner's regulation and performance before, during and after the problem solving. In fact, the metacognitive questioning mediates between the task, the student interactions and the cognitive responses by directing learners to provide elaborated explanations in response to the metacognitive questioning (King, 1998).

For example, in the ASK to THINK – TEL WHY method (King, 1998) students learn in pairs, where one peer is the tutor and the other is the tutee. The tutor asks five types of "why" and "how" questions, instead of "what" questions. These consist of review questions ("what does … mean? Describe in your own words"); thinking questions ("what is the difference between … and …?" "what do you think would happen if …?"); probing questions as needed ("please elaborate"); hint questions ("have you thought about …?" "how can … help you?"); and metacognitive questions ("what did you learn that you did not know before?" "how will you remember this?") The tutee's answers boosts further metacognitive and cognitive interactions between the dyad members. Hence, the mutual questioning facilitates the scaffolding of each peer-member's thinking.

Lai (2011) describes several teaching techniques aimed at improving mathematical performance and metacognition in the classroom. The primary techniques include, but are not limited to, thinking aloud, discussing and articulating, using checklists, self-questioning, teaching strategies, modelling, and providing feedback. These techniques have been commonly applied in both experimental studies and in the classroom (e.g. Gama, 2004; Lai, 2011), and are usually integrated in various metacognitive pedagogical models.

Conclusion

In almost all fields, talented people intensively practise, often with the aid of a mentor who explicitly guides them in planning, monitoring, controlling and evaluating their accomplishments. Ordinary students also have to be explicitly guided in how to activate metacognitive processes, followed by intensive practice embedded within the learned content (domain specific). The need to include metacognitive scaffolding has been evident in various learning environments and different age groups, as explained in the next chapters.

Metacognition can be taught and should be intensively practised in mathematics classrooms as well as in other disciplines. Improving metacognitive skills has positive benefits for academic achievement, particularly for CUN problem solving.

Co-operative learning methods have the potential to enhance cognitive and metacognitive processes because such methods provide ample opportunities for students to articulate their thinking, use mathematics language, work within their zone of proximal development, provide elaborated explanations, and be involved in conflict resolutions and mutual learning. To fulfil this potential, students still need to be guided in how to apply metacognition in their learning.

Metacognitive pedagogy has to be embedded in the subject content matter and explicitly taught. Successful mathematics students are metacognitively active. They think what the problem is all about, compare the problem in hand with problems solved in the past to find similarities and differences, suggest strategies that are appropriate for solving the problem, and reflect on all stages of the solution process. All these activities are key elements in metacognitive pedagogies.

It would be useful to inform teachers and students on the contributions of metacognition to the learning processes, and in particular to the solution of CUN problems.

References

Antil et al. (1998), "Cooperative learning: Prevalence, conceptualizations, and the relation between research and practice", *American Educational Research Journal,* Vol. 35(3), pp. 419-454.

Aronson, E. and S. Patnoe (2011), *Cooperation in the Classroom: The Jigsaw Method* (3rd edition), Pinter and Martin Ltd, London.

Artzt, A.F. and C.M. Newman (1990), "Cooperative learning", *Mathematics Teacher,* Vol. 83, pp. 448-449.

Artzt, A.F. and S. Yaloz-Femia (1999), "Mathematical reasoning during small-group problem solving", in L. Stiff and F. Curio (eds.), *Developing Mathematical Reasoning in Grades K–12: 1999 Yearbook,* National Council of Teachers of Mathematics, Reston, VA.

Bloom, B.S. (ed.) (1985), *Developing Talent in Young People,* Ballantine Books, New York.

Cardelle-Elawar, M. (1995), "Effects of metacognitive instruction on low achievers in mathematics problems", *Teaching and Teacher Education,* Vol. 11(1), pp. 81-95.

Dansereau, D.F. (1988), "Cooperative learning strategies", in C.E. Weinstein, E.T. Goetz and P.A. Alexander (eds.), *Learning and Study Strategies: Issues in Assessment, Instruction, and Evaluation,* Academic Press, Orlando, FL, pp. 103-120.

De Corte, E., L. Verschaffel and P. Op 't Eynde (2000), "Self-regulation: A characteristic and a goal of mathematics education", in M. Bockaerts, P.R. Pintrich and M. Zeidner (eds.), *Handbook of Self-Regulation,* Academic Press, San Diego, CA, pp. 687-726.

Desoete, A. and M.V.J. Veenman (2006), "Introduction", in A. Desoete and M.V.J. Veenman (eds.), *Metacognition in Mathematics Education,* Nova Science Publishers, New York, pp. 1-10.

Dignath, C. (2012), "What teachers think about self-regulated learning (SRL). An investigation of teachers' knowledge and attitude towards SRL and their effects on teacher instruction of SRL in the classroom", in *Metacognition 2012 – Proceedings of the 5th Biennial Meeting of the EARLI Special Interest Group 16 Metacognition,* Milan, 5-8 September.

Dignath, C. and G. Buettner (2008), "Components of fostering self-regulated learning among students. A meta-analysis on intervention studies at primary and secondary school level", *Metacognition Learning,* Vol. 3, pp. 231-264.

Dignath, C., G. Buettner and H.-P. Langfeldt (2008), "How can primary school students learn self-regulated learning strategies most effectively? A meta-analysis on self-regulation training programmes, *Educational Research Review,* Vol. 3(2), pp. 101-129.

Efklides, A. (2008), "Metacognition: Defining its facets and levels of functioning in relation to self-regulation and co-regulation", *European Psychologist,* Vol. 13(4), pp. 277-287.

Fantuzzo, J. et al. (2007), "Investigation of dimensions of social-emotional classroom behavior and school readiness for low-income urban preschool children", *School Psychology Review*, Vol. 36(1), pp. 44-62.

Fuchs et al. (2001) "Oral reading fluency as an indicator of reading competence: A theoretical, empirical, and historical analysis", *Scientific Studies of Reading*, Vol. 5(3), pp. 239-256.

Gama, C.A. (2004), "Integrating metacognition instruction in interactive learning environments", PhD Thesis, University of Sussex, http://homes.dcc.ufba.br/~claudiag/thesis/Thesis_Gama.pdf.

Hadwin, A.F., S. Järvelä and M. Miller (2011), "Self-regulated, co-regulated, and socially-shared regulation of learning", in B.J. Zimmerman and D.H. Schunk (eds.), *Handbook of Self-Regulation of Learning and Performance*, Routledge, New York, pp. 65-84.

Hattie, J.A. (1992), "Measuring the effects of schooling", *Australian Journal of Education*, Vol. 36(1), pp. 5-13.

Hattie, J.A., J. Biggs and N. Purdie (1996), "Effects of learning skills interventions on student learning: A meta-analysis", *Review of Educational Research*, Vol. 66(2), pp. 99-136.

Hoppenbrouwers, S. and H. Weigand (2000), "Meta-communication in the language action perspective", in the *Proceedings of the Fifth International Workshop on the Language-Action Perspective on Communication Modelling* (LAP 2000), Aachen, Germany, 14-16 September.

Johnson, D.W. and R.T. Johnson (1999), *Learning Together and Alone: Cooperative, Competitive, and Individualistic Learning* (5th edition), Prentice-Hall, Englewood Cliffs, N.J.

King, A. (1998), "Transactive peer tutoring: Distributing cognition and metacognition", *Educational Psychology Review*, Vol. 10(1), pp. 57-74.

Kistner, S. et al. (2010), "Promotion of self-regulated learning in classrooms: Investigating frequency, quality, and consequences for student performance", *Metacognition Learning*, Vol. 5, pp. 157-171.

Kitsantas, A. and M. Kavussanu (2011), "Acquisition of sport knowledge and skill: The role of self-regulatory processes", in B.J. Zimmerman and D. Schunk (eds.), *Handbook of Self-Regulation of Learning and Performance*. Routledge, New York, pp. 217-233.

Kramarski, B. and V. Dudai (2009), "Group-metacognitive support for online inquiry in mathematics with differential self-questioning", *Journal of Educational Computing Research*. Vol. 40(4), pp. 365-392.

Kramarski, B., Z.R. Mevarech and M. Arami (2002), "The effects of metacognitive training on solving mathematical authentic tasks", *Educational Studies in Mathematics,* Vol. 49, pp. 225-250.

Kramarski, B. and N. Mizrachi (2006), "Online discussion and self-regulated learning: Effects of instructional methods on mathematical literacy", *The Journal of Educational Research,* Vol. 99(4), pp. 218-230.

Kramarski, B. and S. Zoldan (2008), "Using errors as springboards for enhancing mathematical reasoning with three metacognitive approaches", *The Journal of Educational Research,* Vol.102(2), pp. 137-151.

Kuhn, D. and D. Dean (2004), "Metacognition: A bridge between cognitive psychology and educational practice", *Theory into Practice,* Vol. 43(4), pp. 268-273.

Lai, R.E. (2011), "Metacognition: A literature review", Pearsons Research Report, *www.pearsonassessments.com/hai/images/tmrs/Metacognition_Literature_Review.*

Lin, X. (2001), "Designing metacognitive activities", *Educational Technology Research and Development,* Vol. 49(2), pp. 23-40.

Marzano, R.J. (1998), *A Theory-Based Meta-Analysis of Research on Instruction,* Mid-Continental Regional Educational Laboratory, Aurora, CO.

McClain, K. and P. Cobb (2001), "Supporting students' ability to reason about data", *Educational Studies in Mathematics,* Vol. 45, pp. 103-129.

McLeod, L. (1997), "Young children and metacognition: Do we know what they know they know? And if so, what do we do about it?", *Australian Journal of Early Childhood,* Vol. 22(2), pp. 6-11.

Mevarech, Z.R. (1991), "Learning mathematics in different mastery environments", *Journal of Educational Research,* Vol. 84(4), pp. 225-231.

Mevarech, Z.R. (1985), "The effects of cooperative mastery learning strategies on mathematics achievement", *Journal of Educational Research,* Vol. 78(6), pp. 372-377.

Mevarech, Z.R. and B. Kramarski (1997), "IMPROVE: A multidimensional method for teaching mathematics in heterogeneous classrooms", *American Educational Research Journal,* Vol. 34(2), pp. 365-395.

Mevarech, Z.R. and P. Light (1992), "Peer-based interaction at the computer: Looking backward, looking forward", *Learning and Instruction,* Vol. 2(3), pp. 275-280.

Mevarech, Z.R. and G. Shabtay (2012), "Judgment-of-learning and confidence in mathematics problem solving: A metacognitive benefit for the explainer", in *Metacognition 2012 – Proceedings of the 5th Biennial Meeting of the EARLI Special Interest Group 16 Metacognition,* Milan, 5-8 September.

Palincsar, A.S. and A. Brown (1984), "Reciprocal teaching of comprehension-fostering and comprehension-monitoring activities", *Cognition and Instruction,* Vol. 1(2), pp. 117-175.

Piaget, J. (1985), *The Equilibration of Cognitive Structures: The Central Problem of Intellectual Development*, The University of Chicago Press, Chicago.

Sangster Jokic, C. and D. Whitebread (2011), "The role of self-regulatory and metacognitive competence in the motor performance difficulties of children with developmental coordination disorder: A theoretical and empirical review", *Educational Psychology Review*, Vol. 23, pp. 75-98.

Schoenfeld, A.H. (1992), "Learning to think mathematically: Problem solving, metacognition, and sense-making in mathematics", in D.A. Grouws (ed.), *Handbook of Research on Mathematics Teaching*, MacMillan Publishing, New York, pp. 334-370.

Schraw, G., K.J. Crippen and K. Hartley (2006), "Promoting self-regulation in science education: Metacognition as part of a broader perspective on learning", *Research in Science Education*, Vol. 36, pp. 111-139.

Schraw, G. and D. Moshman (1995), "Metacognitive theories", *Educational Psychology Review*, Vol. 7(4), pp. 351-371.

Sharan, Y. and S. Sharan (1992), *Expanding Cooperative Learning through Group Investigation*, Teachers College Press, New York.

Slavin, R.E. (2010), "Co-operative learning: What makes group-work work?", in H. Dumont, D. Istance and F. Benavides (eds.), *The Nature of Learning: Using Research to Inspire Practice*, Educational Research and Innovation, OECD Publishing, Paris, http://dx.doi.org/10.1787/9789264086487-9-en.

Slavin, R.E. (1994), *Using Student Team Learning*, 3rd edition, Johns Hopkins University, Baltimore, MD.

Slavin, R.E. and C. Lake (2008), "Effective programs in elementary school mathematics: A best-evidence synthesis", *Review of Educational Research*, Vol. 78(3), pp. 427-515.

Slavin, R.E., C. Lake, and C. Groff (2009). "Effective programs in middle and high school mathematics: A best-evidence synthesis", *Review of Educational Research*, Vol. 79(2), pp. 839-911.

Steen, G.J. (1999), "Genres of discourse and the definition of literature", *Discourse Processes*, Vol. 28(2), pp. 109-120.

Stillman, G. and Z.R. Mevarech (2010), "Metacognition research in mathematics education: From hot topic to mature field", *ZDM International Journal on Mathematics Education*, Vol. 42(2), pp. 145-148.

Veenman, M.V.J., B.H.A.M. Van Hout-Wolters and P. Afflerbach (2006), "Metacognition and learning: Conceptual and methodological considerations", *Metacognition and Learning*, Vol. 1, pp. 3-14.

Verschaffel, L., F. Depaepe and E. De Corte (2007), "Upper elementary school teachers' conceptions about and approaches towards mathematical modelling and problem solving: How do they cope with reality?", Paper presented at the Conference on Professional Development of Mathematics Teachers Research and Practice from an International Perspective held at the Mathematische Forschungsinstitut Oberwolfach, Germany.

Vygotsky, L.S. (1978), *Mind in Society: The Development of Higher Psychological Processes*, Harvard University Press, Cambridge, MA.

Webb, N.M. (2008), "Learning in small groups", in T.L. Good (ed.), *21st Century Education: A Reference Handbook,* Sage Publications, Los Angeles, pp. 203-211.

Zion, M., T. Michalsky and Z.R. Mevarech (2005), "The effects of metacognitive instruction embedded within an asynchronous learning network on scientific inquiry skills", *International Journal of Science Education,* Vol. 27(8), pp. 957-958.

Chapter 4

Metacognitive pedagogies in mathematics education

This chapter reviews the five main metacognitive pedagogies used in maths education, their benefits and trade-offs. The models are: Polya, Schoenfeld, IMPROVE, Verschaffel and Singapore. All of them use some form of self-directed questions but differ in their details, scope and age range. Polya's and Schoenfeld's models are designed to be used with university students and on single CUN problems, whereas IMPROVE, Verschaffel's model and the Singapore model can be used with younger learners and for a set of problems or even a whole curriculum. IMPROVE has also been modified for use in other domains, and for teachers' professional development with or with no advanced technologies. Comparing the models highlights the advantages and challenges associated with each one of them.

Teachers might think at first glance that they already spontaneously apply metacognitive instruction in teaching, or that students automatically implement metacognitive strategies in learning. Observations have shown, however, that this is rarely the case. Teachers often implement metacognitive strategies in implicit ways and do not devote the time to explain to students the importance of the metacognitive processes and how to implement them (Dignath and Buettner, 2008; Dignath et al., 2008).

Mathematics teachers confront additional challenges. Many mathematics teachers focus solely on the mathematics itself, believing that everything else is either not important, or not on their agenda. Consequently, maths teachers emphasise the practising of cognitive skills, but rarely do the same with regards to metacognitive processes. Our experience shows that when teachers experience the benefits of metacognitive instruction, these approaches change.

The inclusion of complex, unfamiliar, and non-routine (CUN) tasks in the mathematics curriculum, and the strong relationships between metacognition and school attainment have further increased the importance of training learners to monitor, control and evaluate the problem-solving processes. Polya (1957), Schoenfeld (1985), Mevarech and Kramarski (1997), Verschaffel (1999), and the Singapore National Institute of Education (Lianghou and Yan, 2007) developed pedagogical models of metacognitive instruction for a variety of age groups. We chose to review these interventions because they have been used either for research or practical purposes, and all except Singapore reported the advantages and trade-offs with regards to mathematics achievements. Singapore is the only country in which metacognition is part of the mathematics curriculum and implemented nationwide. Interestingly, Singapore is also the country that tops the "creative problem solving" international test and that has consistently scored high in international tests of learning outcomes (OECD, 2013, 2014).

Over the years, other metacognitive pedagogies have been developed and evaluated (e.g. Garofalo and Lester, 1985). The models of Polya, Schoenfled, Mevarech and Kramarski, Verschaffel, and Singapore provide the basis for metacognitive pedagogies in mathematics education, particularly those that mainly focus on CUN tasks.

Polya's heuristics for solving maths problems

Polya (1949), a well-known mathematician, proposed a four-stage problem-solving model, called "How to Solve it?". Even though Polya did not use the terms associated with metacognition, which were only introduced in the late 1970s, his model and heuristics actually refer to what we currently conceptualise as metacognition (Figure 4.1).

Polya (1949) based his model on the notion of heuristics, which are defined as "a rule of thumb for making progress on different problems" (Schoenfeld, 1985). In Polya's model there are heuristics for every stage: *understanding* (identifying the givens, wanted and conditions), *devising a plan* (making connections to existing knowledge), *carrying out the plan* (checking each step), and looking backward (checking the result and looking for alternative ways of solution).

Figure 4.1. **Polya's four-stage model**

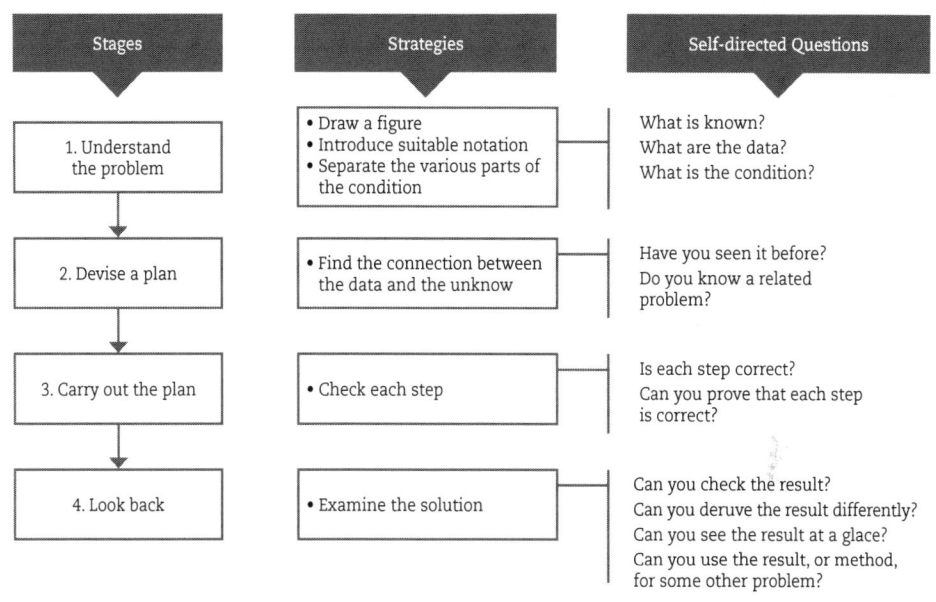

Source: Polya, G. (1949), *How to Solve It*, 1ˢᵗ Edition, Princeton University Press, Princeton, NJ. Based on Gama (2004) presentation.

Polya's model was adopted by mathematicians right away. Although these heuristics had never been taught as such, many mathematicians felt Polya had successfully opened the "black box", exactly describing what they are doing in solving mathematics problems (Schoenfeld, 1985). Maths experts and maths educators shared the idea that using these self-directed questions was essential to the problem-solving process (Schoenfeld, 1985).

Over time, the model became recognised all over the world. Mathematicians, maths educators, scientists, and people engaged in solving other kinds of problems, including routine and CUN problems, felt that Polya's model was right, because of its high face validity, i.e. it looked like it should work.

Yet, these high expectations resulted in disappointment. When maths educators tried to apply the model in the classrooms it simply did not work. As Schoenfeld (1987) wrote, "there was empirical evidence to suggest that something was either wrong or missing... Despite the enthusiasm for the approach, there was no clear evidence that the students had actually learned more as a result of their heuristic instruction, or that they had learned any general problem solving skills that transferred to novel situations" (p. 288). Schoenfeld summarised his observations saying "at a certain level, Polya's descriptions of problem solving strategies were right. If you already knew how to use strategies, you recognized them in his writing. But at a finer grain size, Polya problem solving descriptions did not contain enough detail for people unfamiliar with the strategies to be able to implement them" (p. 288).

Schoenfeld's metacognitive instructional model

Being fascinated by Polya's model on the one hand, and recognising its weaknesses on the other, Schoenfeld (1985) proposed a problem-solving model consisting of the following stages:

- analysis, oriented toward understanding the problem by constructing an adequate representation;

- design of a global solution plan;

- exploration, oriented toward transforming the problem into a routine task;

- implementation of the solution plan;

- verification of the solution.

To enhance the use of these processes, Schoenfeld suggests applying a set of three self-directed questions:

- What exactly are you doing? (Can you describe it precisely?)

- Why are you doing it? (How does it fit into the solution?)

- How does it help you? (What will you do with the outcome when you obtain it?)

Addressing these questions serves two purposes: first, it encourages students to articulate their problem-solving strategies; and second, it induces reflections on those activities. Training students to spontaneously ask themselves those questions might lead them to think about their thinking, and regulate and monitor their own cognitive processes.

In Schoenfeld's model, the different stages are performed consecutively, and the corresponding relevant heuristics are explained and practised. The model is extensively used to demonstrate how experts select and apply the heuristics. The instructor exemplifies the use of the self-directed metacognitive questions and students spent one-third of the course time solving problems in small groups by following the stages described above. While students practise the method, the instructor takes the role of a consultant, and provides external regulation in the forms of hints, prompts or feedback.

Schoenfeld (1985, 1989, 1992) applied his metacognitive instructional model in teaching unfamiliar mathematics problems to students in Stanford University. Students were given 20 minutes to solve the problem and their activities during the 20 minutes are presented in Figure 4.2 below. Prior to the intervention, students spent less than five minutes reading the problem and the rest of the time exploring the solution, whereas after the intervention students applied various metacognitive strategies, including planning, exploring and verification; they tried to implement the suggested solution, reflected on the outcome, and tried again. Schoenfeld further reported that prior to the intervention about 60% of the students tended to read the problem and quickly choose a solution strategy. They then pursued this strategy, even if they had clear evidence that they were not making progress. In contrast, by the end of the training, fewer than 20% of the students followed the original "jump into one solution attempt and pursue it no matter what" approach (Schoenfeld, 1992). Applying the self-directed questioning described above, resulted in better solutions to the CUN problems. Using the metacognitive prompt "Can you describe exactly what are you doing?" led students to carefully plan the solution; the questions "Why are you doing it?", and "How will it help you?" guided them to implement their plan and think how the outcome would help them in the next step. The most important observation is that after the intervention, the students were not constrained by their first attempt; instead they were flexible, trying different approaches when they obtained sub-solutions that did not fit into the solution.

Figure 4.2. Solving a problem with and without self-directed questioning: timeline of activities

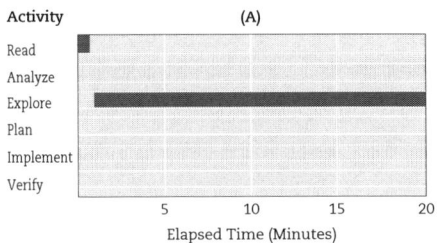

Time-line graph of a typical student attempt to solve a non-standard problem.

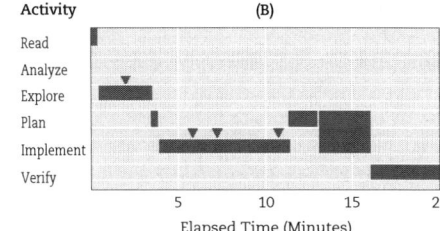

Time-line graph of two students working a problem after the problem solving course.

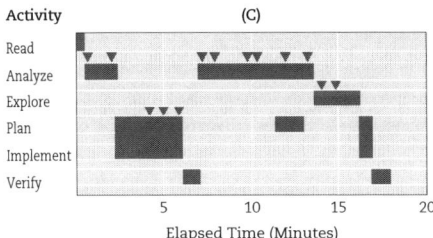

Time-line graph of a mathematician working a difficult problem.

Source: Schoenfeld, A.H. (1985), *Mathematical Problem Solving.* Academic Press, New York, p. 37.

Applying the self-directed questioning described above resulted in better solutions of the CUN problem.

Schoenfeld's model was implemented mainly at the university level with students who majored in mathematics. The self-addressed questions described above might be suitable for university students who can speculate on how each step leads to the solution of the next step, and hypothesise what they will do with the outcome when they will obtain it (Questions 2 and 3). However, these questions might create cognitive overload in younger students and also be too theoretical for routine problems. Both Schoenfeld's and Polya's models had to be reconstructed to be used by younger students for whom mathematics is compulsory, and who have to be explicitly guided in regulating the solution processes of both routine and CUN tasks.

IMPROVE model

One of the first metacognitive instructional methods for primary to secondary school students is IMPROVE, designed by Mevarech and Kramarski (1997). The acronym IMPROVE describes the teaching stages that constitute the method:

- *Introducing* the whole class to the new material, concepts, problems or procedures by modelling the activation of metacognitive processes.

- *Metacognitive* self-directed questioning applied in small groups or individualised settings.

- *Practising* by employing the metacognitive questioning.

- *Reviewing* the new materials by the teacher and the students, using the metacognitive questioning.

- *Obtaining* mastery on higher and lower cognitive processes.

- *Verifying* the acquisition of cognitive and metacognitive skills based on the use of feedback-corrective processes.

- *Enrichment* and remedial activities.

The big challenge for Mevarech and Kramarski was to design an innovative instructional method that could be administered by ordinary teachers in "regular" mathematics classrooms, which often include a large number of students of varying mathematical abilities – some experiencing deep difficulties in mathematics, with others having an excellent record of mathematics achievement.

Mevarech and Kramarski (1997) also took the pedagogies suggested by Polya (1949) and Schoenfeld (1985, 1989, 1992) one step further by proposing a model that focuses not only on the teaching of a single CUN problem as Polya and Schoenfeld did, but also on the teaching/learning of an entire session, a complete unit, or the whole maths curriculum, including the provision of enrichment and remedial activities. The IMPROVE learning environment includes a variety of carefully designed materials and challenging problems to be solved either in co-operative or individualised settings, with or without ICT. The metacognitive guidance is embedded within the ongoing teaching/learning processes, rather than being considered as a "nice to have" supplement. Metacognition, both knowledge and regulation, is the central component in each stage of IMPROVE.

IMPROVE is rooted in various paradigms: cognition, social cognition and self-regulated learning. It is unique in its synergetic approach, integrating various theories into one generic entity that can be applied in learning different kinds of tasks (routine, CUN, or even coping with emotional problems), various contexts, and in immediate and delayed assessments.

The key element of IMPROVE is the use of four types of self-directed metacognitive questioning based on Polya's (1957) and Schoenfeld's (1989) studies:

- Comprehension questions: what is the problem all about?

- Connection questions: how is the problem at hand similar to, or different from problems I have already solved? Please explain your reasoning.

- Strategic questions: what kinds of strategies are appropriate for solving the problem, and why? Please explain your reasoning.

- Reflection questions: does the solution make sense? Can the problem be solved in a different way? Am I stuck? Why?

This series of questions guides the learner to activate metacognitive processes before, during and at the end of solving the problem. Applying it might become a mental habit, allowing students to use it not only in mathematics, but also in other problem-solving situations and during life-long learning.

Comprehension questions

Observing students solving either typical or complex problems, teachers and researchers see that quite often students immediately start "solving" the problem without attempting to comprehend the problem they have to solve (e.g. Schoenfeld, 1992). Frequently, students (mistakenly) rely on the shallow "story" of the given word problem rather than on its mathematical construct, or (mistakenly) focus on the problem's keywords. For example, students mistakenly assume that "more" always implies addition, even when this is not the case. In the problem: a t-shirt costs EUR 10 in Store A which is EUR 2 more than in Store B. How much does the t-shirt cost in Store B? Many students answer EUR 12 rather than EUR 8 (based on Mevarech, 1999).

Obviously, comprehending the problem is the first stage in the solution process. The comprehension question guides students to think what the problem is all about. An effective way to address the comprehension question is to ask students to "tell" the problem in their own words rather than rereading it, or to identify what kind the given problem is (e.g. this is a speed-time-distance problem), without referring to the specific numbers mentioned in the problem.

Connection questions

Current theories in cognitive psychology assume that knowledge is constructed by making connections (Wittrock, 1986). Without the construction of bridges between existing and new knowledge, the new information remains discrete and innate (King, 1991). Hence, prior knowledge is a cornerstone in the learning process (Schneider and Stern, 2010). The connection question guides students to construct those bridges by asking themselves "how the problem at hand is similar to or different from what I already know or from problems I have already solved?" When students address the connection question, they are less likely to use trial and error which can result in failure, frustration and a tendency to avoid mathematics (Schoenfeld, 1992).

Strategic questions

The dictionary definition of strategy is "a plan, method, or series of manoeuvres or stratagems for obtaining a specific goal or result; skilful use of a stratagem, for example: the salesperson's strategy was to seem always to agree with the customers". In IMPROVE students are trained to use two kinds of strategies: mathematics strategies, and metacognitive strategies. Box 4.1 provides a list of cognitive and metacognitive strategies focusing on mathematics problem solving.

Such a large repertoire of strategies raises the question of how teachers could possibly teach them all. Obviously, students do not have to memorise all these strategies, nor do they have to acquire them in one session or in one year. Strategy acquisition continues throughout life, in school and out of school, during learning and at work (Lave, 1988; Nunes, Schliemann and Carraher, 1993).

Modelling through thinking-aloud is one of the best ways to make students aware of the strategies. When teachers explicitly label the strategies in a way that describes their meaning and show students how to use them, and when students practice them, students gradually gain a rich repertoire of strategies. At a certain stage, applying both kinds of strategies, cognitive and metacognitive, may become automatic.

Reflection questions

The purpose of the reflection questions is threefold: 1) guide students in monitoring their progress as they solve problems; 2) assist students in making changes and adapting their strategies when they are "stuck"; and 3) direct students to look back and analyse what works and how can they use it in solving other problems, or to think of alternative ways (e.g. more "elegant" or quicker) of solving the problem. In IMPROVE the reflection questions are as follows:

Ask yourself:

• Does the solution make sense? Does it fit the conditions described in the problem? How many solutions should I get?

Box 4.1. Cognitive and metacognitive mathematics strategies

This box briefly describes some of the cognitive and metacognitive strategies used in solving different kinds of mathematics problems. The list is mainly based on Google Mathematics and Science Strategies: Professional Development Resource – Teacher V, Kujawa and Huske (1995) and the studies reviewed in this manuscript.

Cognitive strategies for solving mathematics problems:

- classifications that are thoughtfully labeled to indicate the characteristics of the task
- comparisons of items, groups, or quantities
- using manipulations to help representations
- guess and check
- make a table
- make a drawing/picture
- systematically eliminate possible hypotheses/procedures/theorems to use
- use a formula
- find a pattern and use models to describe patterns
- simplify the problem by looking at specific cases (e.g. what if x=0)
- divide a complex problem into simpler problems and solve each one separately
- estimate the answer before making the calculations and then check if the calculated answer is close to your initial estimation
- use number sense
- work backwards
- distinguish between relevant and irrelevant information
- identify the givens and the wanted and check if you used all the givens
- make generalisations about numbers
- use various techniques to display the data
- use reading strategies in comprehending CUN and word problems

Developing a plan of action, maintaining/monitoring the plan, and evaluation

Before you develop the plan of action ask yourself:
- What in my prior knowledge will help me with this particular task?
- Have I solved problems like that before? How?
- What strategies work best for me (visualising, writing, memorising, diagraming, self-testing etc.)?
- In what direction do I want to go?
- What should I do first?
- How much time do I have to complete the task?
- What is my goal / how motivated am I? This question aims to enhance motivation and reminds students that without it they will not succeed.

During the problem-solving process when you maintain/ monitor the plan of action ask yourself:
- How am I doing?
- What am I doing here? Why am I doing it? Am I on the right track?
- How should I proceed?
- What information is relevant or important to remember/consider/use?
- Should I move in a different direction?
- Should I adjust the pace depending on the difficulties?
- Am I stuck? Why? Did I refer to all the relevant information? (systematically go over all the information and evaluate if you considered all the givens and the wanted)
- What do I need to do if I do not understand?

At the end of the solution process:

When you are evaluating the plan of action ask yourself:
- Does the solution make sense? Does it fit the information given in the problem?
- How well did I do?
- Did my particular course of thinking produce more or less than I had expected?
- What could I have done differently? (even when the answer is correct)
- How might I apply this line of thinking to other problems?
- Do I need to go back though the task to fill in any "blanks" in my understanding?

Source: Google Mathematics and Science Strategies: Professional Development Resource – Teacher V ; Kujawa and Huske (1995), *Strategic Teaching and Reading Project Guidebook*, NCREL (North Central Regional Educational Laboratory).

- Can I solve the problem differently? Can I solve it in a more "elegant" way, or in a shorter way? How?

- How can I use what I have learned now in solving other problems?

- Am I stuck? Why am I stuck? Did I consider all the information given in the problem? Did I identify correctly all the givens and the wanted?

IMPROVE is an ecologically valid method that approximates the classroom situation by referring to all the teaching stages from the introduction of a new topic, concept, or problem through the assessment and up to the stage of provision of remedial and enrichment activities (i.e. the last stage of IMPROVE, see below). The four generic metacognitive self-directed questions are easy to remember and use, and the learning in small groups further facilitates the implementation of metacognitive processes during the articulation of one's thinking.

Verschaffel's model of metacognitive instruction for upper elementary school maths

Other researchers have also proposed using heuristics and metacognition in solving mathematics problems. One of them is Verschaffel (1999) who developed a broader model for solving routine and non-routine problems for upper elementary school classrooms. As with the other models described above, Verschaffel's model includes the stages of understanding the problem, planning, executing the plan, interpreting the outcome and formulating an answer. Verschaffel complements the model by describing the specific heuristics for each step (De Corte, Verschaffel and Eynde, 2000, p. 714):

- Build a mental representation of the problem

  Heuristics: draw a picture, make a list, a scheme or a table, distinguish relevant from irrelevant data, and use your real-world knowledge

- Decide how to solve the problem

  Heuristics: make a flowchart, guess and check, look for a pattern simplify the numbers

- Execute the necessary calculations

- Interpret the outcome and formulate an answer

- Evaluate the solution

Similar to IMPROVE, a lesson using Verschaffel's model also consists of small-group problem solving activities or individual assignments, always followed by a whole-class discussion. Each metacognitive strategy is initially demonstrated by the teacher, whose role is to encourage students to engage in mathematics problem solving and to reflect upon the kinds of cognitive and metacognitive activities involved in the process. These encouragements and scaffolds are gradually withdrawn as the students become more competent and take more responsibility for their own learning and problem solving.

Singapore model of mathematics problem solving

Singapore, rated as one of the top-achieving countries by both the Trends in International Mathematics and Science Study (TIMSS) and the OECD Programme for International Student Assessment (PISA), adopted the concept of metacognition in its mathematics curriculum for all school grade levels at the start of the 2000s (Lianghu and Yan, 2007). The mathematical problem-solving framework (see Figure 4.3), combines five inter-related components: 1) concepts (numerical, geometrical, algebraic, statistical, probabilistic and analytical); 2) skills (numerical calculation, algebraic manipulation, spatial visualisation, data analysis, measurement, use of mathematical tools and estimation); 3) processes (reasoning, communication and connections, thinking skills and heuristic, application and modelling); 4) metacognition (monitoring one's own thinking, self-regulation of learning); and 5) attitudes (beliefs, interest, appreciation, confidence and perseverance).

Figure 4.3. **Singapore's pentagonal framework for mathematical problem solving**

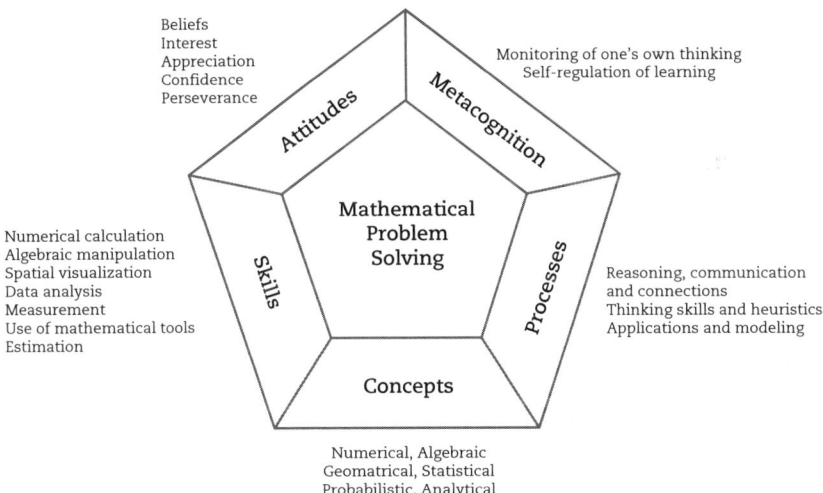

Source: Singapore Ministry of Education.

The pentagon framework has influenced the development of the latest Singapore mathematics textbooks to include routine, non-routine and authentic problems, as well as exploration and project tasks at the end of each chapter (Lianghou and Yan, 2007). Teachers have started to explicitly encourage students to apply self-regulation and self-reflection in mathematical problem solving. The mathematics syllabi included problem-solving heuristics and a model for problem solving, largely following the Polya model:

1. understanding the problem

2. devising a plan (choosing a heuristic)

3. carrying out the plan

4. needs modification / a new plan?

5. checking: does the answer make sense, is the answer reasonable?

6. looking back (reflection): improving on the method used, seeking alternative solutions, and expanding the method to other problems.

In Singapore, metacognitive guidance is commonly implemented nationwide. The teachers model the heuristics and the metacognitive strategies, and the students practice them on a regular basis, in solving routine as well as CUN problems.

Comparing the metacognitive models

What are the similarities and differences between the metacognitive pedagogies described above? This section compares the models of Polya, Schoenfeld, IMPROVE and Verschaffel but not the Singapore model because we could not locate any study showing the effects on mathematics achievement of its metacognitive component. The lack of evidence probably results from the fact that Singapore model as a whole is compulsory without distinguishing between the components.

Undoubtedly, Polya's model provides the foundation for all other metacognitive models described above. Polya was the first to afford a general outline of "how to solve it". His four stages and self-directed questioning are included in all the other models, although sometimes the terms or number of stages are different. Polya is also the mathematician who emphasised the importance of applying heuristics in solving CUN tasks by the careful application of planning, monitoring and reflecting processes.

The models differ, however, in several aspects. First, while both Polya's and Shoenfeld's models refer to the solution of a single problem, usually a CUN task, Verschafell's model and IMPROVE focus on a set of similar routine and non-routine problems, or a whole unit, with IMPROVE proposing covering the whole curriculum. This is not only a quantitative difference (a single problem versus a whole curriculum) because confining the model to a single task limits the role of practice and does not expose students to the benefits of self-regulation and using metacognitive processes in a variety of problems. Because of that, it might also reduce the possibility of students generalising the model, and transferring it to other situations. Furthermore, the positive outcomes reported by Schoenfeld might result from Hawthorne effects (i.e. people's tendency to perform better when they participate in an experiment, due to the attention they receive from the researchers or the very fact that they are participating in an important study) and the extra time and effort devoted to the solution of those specific problems.

Another difference between the models is the intended grade level: while Polya's and Schoenfeld's models have been implemented mainly in tertiary education, Verschaffel's model was implemented with middle and upper elementary school students, and the IMPROVE model with all ages. The wide range of grade levels enabled Verschaffell and IMPROVE to work on routine and CUN tasks, rather than on only one kind of task. In addition, the IMPROVE model's generic self-addressed questioning could be easily modified to foster socio-emotional outcomes in addition to enhancing mathematics achievement (see Chapter 6). Finally, although all models aim to improve metacognition as a means for promoting mathematics problem solving, there are differences in the scope of the studies done on them (e.g. only mathematics versus mathematics, science, and other domains) and the amount and quality of evidence showing the effectiveness of the models in promoting the sets of skills that are useful innovation-driven societies. Chapter 5 provides evidence-based reviews regarding the effects of these metacognitive pedagogies.

Table 4.1. presents a comparison of metacognitive pedagogical methods.

The above comparison highlights the advantages and challenges of each model. First, in spite of the differences between the models, all have similar aims, frameworks, emphases on CUN tasks (both in teaching and in assessment of the outcomes), use of heuristics, and co-operative learning environments. The basic differences lie in the intended educational level, from a limited age group of mainly university students (Polya and Schoenfeld) or upper elementary school pupils (Verschafell) to K-12 and adults (IMPROVE). Widening the focus to all grade levels enabled the IMPROVE model to cover the teaching of a whole unit/curriculum including the implementation of remedial and enrichment activities, whereas other models focused only on one task, and rarely considered delayed and lasting effects. Some models were designed to be used only in maths classrooms (Schoenfeld and Verschaffel), but IMPROVE has also been modified for use in science education. Last but not least, the models differ in the expected outcomes and their assessments, ranging from one or two capabilities (e.g. Schoenfeld) to a long list of outcomes, including: mathematics reasoning, close and far transfer, math discourse, various metacognitive skills, and affective outcomes, such as mathematics anxiety, motivation, or self-esteem (e.g. IMPROVE).

The main drawback of using metacognitive pedagogies relates to the additional time and efforts that might be associated with its application. However, our experience indicates that students often overcome it after a short period of practice. They recognise the benefits of the metacognitive questioning, easily internalise these questions, and use them in solving problems either individually or in small groups.

Table 4.1. **Metacognitive models compared**

	Polya (1949)	Schoenfeld (1985)	Verschaffel (1999)	IMPROVE (1997)
Framework	Phases: Understanding Planning Carrying out Looking back	Phases: Analysis Design Exploration Implementation Verification	Phases: Representation Planning Execution Interpretation Evaluation	Phases: Introduction Metacogntive questioning Practising Reviewing Obtaining mastery Verification Enrichment and remedial
Important aspects	Cognition Metacognition	Cognitive Metacognition Affect Beliefs	Cognition Metacognition Affect Beliefs	Cognition Metacognition Affect: maths anxiety, motivation, and self-esteem
Foci	Single CUN tasks	Single CUN tasks	Sets of complex, non-routine and realistic word problems	Whole units including: routine problems, CUN tasks and authentic problems
Strategies	Heuristics and metacognition	Heuristics and metacognition	Heuristics and metacognition	Heuristicsand metacogntion
Typical teacher behaviour	Teacher encourages whole-class discussions	Teacher is an external regulator, provides prompts and feedback	Teacher models and scaffolds behavior; Scaffolds are gradually withdrawn	Teacher models metacognitive and cognitive strategies: encourages whole class discussions; provides feedback
Addressing the self-directed questions:	Individually and in class discussion	Individually and in class discussion	Individually and in class discussion	Individually and in class discussion; answering orally or in writing
Learning environments:	Individual/co-operative learning	Individual/co-operative learning	Individual/co-operative learning	Individual/co-operative with or without ICT
Grade level	College students	College students	Upper elementary school	All grades and college students ; pre- and in service teachers
Learning materials	No prepared materials	No prepared materials	Lesson plans	Lesson plans and enrichment and remedial material
Domains	Maths	Maths	Maths	Maths, science and pedagogical content knowledge
Evidence-based findings*		Positive effects on: CUN problems, self-regulation and beliefs	Positive effects on: routine and CUN tasks, maths discourse retention, self-regulation, and beliefs	Positive effects on: routine and CUN tasks, maths reasoning, maths creativity,maths discourse, metacogntion and self-regulation, self-efficacy, and judgment of learning
Duration of effects		Immediate effects	Immediate and delayed effects	Immediate, delayed and lasting effects

* For more details see Chapter 5.

Conclusion

While there is a general consensus that in innovation-driven societies it is not sufficient to teach only routine problems, how best to enhance the solving of CUN problems remains an open question. Five metacognitive pedagogic models address this issue for different age groups and different outcomes: some foster only CUN solutions, while others aim to enhance students' capabilities to solve routine and CUN tasks as well as targeting affective outcomes such as reducing anxiety or enhancing motivation. Each of the five models has been implemented in mathematics classrooms; IMPROVE has been implemented in both mathematics and science classes. These models have significant implications for decision makers and the design of effective learning environments.

- There is large agreement at present that the teaching of metacognitive processes is "doable" in ordinary classrooms with "regular" teachers.

- The use of self-addressed metacognitive questioning is a cornerstone of all these models. It guides students to be aware of the tasks' givens and requirements, their personal prior knowledge and capabilities, and the strategies that might be appropriate for solving the problem.

- Generic metacognitive self-addressed questioning focuses on comprehension, connections, strategies and reflection. These questions activate students' planning, monitoring, control and reflection processes. Hence, the generic metacognitive self-addressed questioning can be applied in various subjects (maths, science, reading, foreign language, etc), as well as for fostering routine and CUN tasks.

- The teaching of metacognition is not limited to a certain age group. Metacognitive pedagogies are applicable in kindergartens, primary and high schools, and in tertiary education.

- Although metacognitive pedagogies can be applied at a very young age, there is no indication that starting earlier has additional advantages for learning. This issue merits future research.

References

De Corte, E., L. Verschaffel and P. Op 't Eynde (2000), "Self-regulation: A characteristic and a goal of mathematics education", in M. Bockaerts, P.R. Pintrich and M. Zeidner (eds.), *Handbook of Self-Regulation,* Academic Press, San Diego, CA, pp. 687-726.

Dignath, C. and G. Buettner (2008), "Components of fostering self-regulated learning among students. A meta-analysis on intervention studies at primary and secondary school level", *Metacognition Learning,* Vol. 3, pp. 231-264.

Dignath, C., G. Buettner, and H.-P. Langfeldt (2008), "How can primary school students learn self-regulated learning strategies most effectively? A meta-analysis on self-regulation training programmes", *Educational Research Review,* Vol. 3(2), pp. 101-129.

Garofalo, J. and F. Lester (1985), "Metacognition, cognitive monitoring and mathematical performance", *Journal for Research in Mathematics Education,* Vol. 16(3), pp. 163-176.

King, A. (1991), "Effects of training in strategic questioning on children's problem-solving performance", *Journal of Educational Psychology,* Vol. 83(3), pp. 307-317.

Kujawa, S. and L. Huske (1995), *Strategic Teaching and Reading Project Guidebook,* NCREL (North Central Regional Educational Laboratory).

Lave, J. (1988), *Cognition in Practice: Mind, Mathematics and Culture in Everyday Life,* Cambridge University Press, Cambridge, United Kingdom.

Lianghou, F. and Z. Yan (2007), "Representation of problem-solving procedures: A representative look at China, Singapore, and US mathematics textbooks", *Educational Studies in Mathematics,* Vol. 66(1), pp. 61-75.

Mevarech, Z.R. (1999), "Effects of metacognitive training embedded in cooperative settings on mathematical problem solving", *Journal of Educational Research,* Vol. 92(4), pp. 195-205.

Mevarech, Z.R. and B. Kramarski (1997), "IMPROVE: A multidimensional method for teaching mathematics in heterogeneous classrooms", *American Educational Research Journal,* Vol. 34(2), pp. 365-395.

Nunes, T., A.D. Schliemann and D.W. Carraher (1993), *Street Mathematics and School Mathematics,* Cambridge University Press, New York.

OECD (2014), PISA 2012 *Results: Creative Problem Solving: Students' Skills in Tackling Real-Life Problems* (Volume V), PISA, OECD Publishing. *http://dx.doi. org/10.1787/9789264208070-en.*

OECD (2013), PISA 2012 *Results: What Students Know and Can Do – Student Performance in Mathematics, Reading and Science* (Volume I, Revised edition, February 2014), PISA, OECD Publishing. *http://dx.doi.org/10.1787/9789264201118-en.*

Polya, G. (1957), *How to Solve It,* 2nd Edition, Princeton University Press, Princeton, NJ.

Polya, G. (1949), *How to Solve It,* 1st Edition, Princeton University Press, Princeton, NJ.

Schneider, M. and E. Stern (2010), "The cognitive perspective on learning: Ten cornerstone findings", in H. Dumont, D. Istance and F. Benavides (eds.), *The Nature of Learning: Using Research to Inspire Practice,* Educational Research and Innovation, OECD Publishing, Paris, *http://dx.doi.org/10.1787/9789264086487-5-en.*

Schoenfeld, A.H. (1992), "Learning to think mathematically: Problem solving, metacognition, and sense-making in mathematics", in D.A. Grouws, (ed.), *Handbook of Research on Mathematics Teaching,* MacMillan Publishing, New York, pp. 334-370.

Schoenfeld, A.H. (1989), "Problem solving in context(s)", in R. Charles and E. Silver (eds.), *The Teaching and Assessing of Mathematical Problem Solving,* National Council of Teachers of Mathematics, Reston, VA, pp. 82-92.

Schoenfeld, A.H. (1987), "Polya, problem solving, and education", *Mathematics Magazine,* Vol. 60(5), pp. 283-291.

Schoenfeld, A.H. (1985), *Mathematical Problem Solving,* Academic Press, New York.

Verschaffel, L. (1999), "Realistic mathematical modelling and problem solving in the upper elementary school: Analysis and improvement", in J.H.M Hamers, J.E.H Van Luit and B. Csapo (eds.), Teaching and Learning Thinking Skills, Swets and Zeitlinger, Lisse, pp. 215-240.

Wittrock, M.C. (1986), "Students' thought processes", in M.C. Wittrock (ed.), *Handbook of Research on Teaching,* MacMillan, New York, pp. 297-314.

Chapter 5

The effects of metacognitive instruction on achievement

Understanding the rationale behind a teaching method and accepting the assumptions on which it is based are not enough. Policy makers, educators and even the public at large look for evidence on its effects on the one hand, and on its drawbacks on the other. A large number of experimental and quasi-experimental studies have been carried out into the effects of metacognitive pedagogies on mathematics achievement, always comparing the metacognitive group to a control group that was taught traditionally. Among school children of all ages, metacognitive approaches improve achievement in arithmetic, algebra and geometry, with lasting effects, and positive effects even in high-stakes situations such as matriculation exams. Effects are similar but smaller for college students. Metacognitive approaches were mostly more effective within co-operative settings, although they also improved achievement among individualised settings.

Tens of studies, if not hundreds, have examined the effects of metacognitive interventions on schooling outcomes. The studies vary, however, with regard to school disciplines, grade levels, types of population, and the outcomes being studied. Many studies focused on reading and writing, others on mathematics (mainly standardised word problems), and still others on disciplines such as natural sciences, humanities and even physical education. A wide range of learners' ages have been studied, from pre-kindergarten through primary and secondary schools, to college and older adults. The research literature also includes studies that examined the effects of metacognitive interventions on people suffering from mental illness and students with learning disabilities, and even studied animals' metacognitive behaviours.

Reviewing all these studies goes beyond the scope of this work. Therefore, we have limited our review to only those studies that focused on school and higher education students' mathematics achievements, at solving either standardised or CUN and authentic problems. In order to gain more insight into how metacognitive interventions work in mathematics classrooms, and given the large number of studies that focused on IMPROVE, this chapter reports the evidence-based findings of IMPROVE in kindergarten, primary, secondary and post-secondary education, followed by reviews of the effects of other metacognitive programmes, and summary studies that used a meta-analysis approach where relevant meta-analysis studies were available.

The impact of metacognitive programmes on problem solving across age groups

This section considers the evidence for the impact of metacognitive pedagogies on kindergarten children, primary and secondary school students, and college students. Within each level of education, the section reviews studies that focused on different kinds of skills, including routine and CUN test scores, reasoning, and other higher order thinking (e.g. transfer of knowledge to new domains). In addition, it addresses the issue of whether metacognitive pedagogies also help students pass high-stake exams.

Kindergarten children

Unfortunately, only a small number of studies have focused on the effects of metacognitive interventions on mathematical thinking of kindergarten children. The literature is flooded with tips on how to promote metacognitive skills in kindergarten mainly by encouraging children to articulate their thinking, but rigorous examinations of the effectiveness of those tips are rather rare.

Despite the disagreement about the age in which children can activate metacognitive processes (see Chapter 2) it is largely believed that the early years (preschools and primary schools) are important for the development of metacognitive skills (Anderson, 2002; Blair, 2002). Recent studies in the area of neuroscience show that much of the brain in young children is plastic, being shaped by experience during the early years of life (e.g. Hinton and Fischer, 2010). Based on a large number of studies, Hinton and Fischer (2010) conclude "Nature and nurture continuously interact to shape brain development. Though certain genetic predispositions exist, the environment powerfully influences how the brain develops. It is therefore often possible and desirable to shift policy from a focus on treating the individual toward a focus on restructuring the environment" (p. 127)… "Learning environments can be structured to build on young children's biological inclination to understand the world numerically and their informal knowledge base to facilitate their understanding of formal mathematics" (p. 128).

Indeed, Whitebread and Coltman (2010) reported evidence showing metacognitive discourse among children of three to five years of age who participated in natural settings in UK kindergartens. They showed how metacognitive pedagogical interactions encouraged children to articulate their thinking, which in turn support metacognitive and self-regulated mathematics behaviour. Recently, Alin (2012) describes similar findings regarding the positive effects of teaching linear measurements in kindergarten by using metacognitive instruction.

Based on these encouraging findings regarding the capability of young children to activate metacognitive processes (Whitebread et al., 2009; Whitebread and Coltman, 2010), Mevarech and Hillel (2012) modified IMPROVE to implement it with children aged four to five years old attending kindergartens. In this study, the mathematical unit focused on division by two, and the metacognitive skill was planning, which is considered to be one of the more difficult and late-developing competencies (Schraw and Moshman, 1995). IMPROVE children were trained to plan their maths activities ahead of time and to articulate their reasoning; the control group executed the same mathematical activities for the same duration of time with no explicit metacognitive intervention. Findings indicate that IMPROVE children were better able to plan ahead, better able to generalise the mathematical principle about division by two of even and odd numbers, and they could also justify their reasoning more accurately than the control group who were not trained to articulate their reasoning.

A related study by Neeman and Kramarski (submitted) examined the effects of IMPROVE on kindergarten children's mathematics problem solving, metacognition and social communication while children worked in small groups on a task that required them to find a pattern. The children's behaviour was compared to that of a control group which was not exposed to metacognitive scaffolding. Findings indicate that children exposed to the IMPROVE model developed a higher level of mathematical problem solving, metacognitive processes and self-efficacy compared with children from the control group. Within the IMPROVE group, children displayed richer explanations, metacognitive expressions and verbal interactions with other peers in the group, including appraisals of their peers' solutions and correcting mistakes. In contrast, communication within the control group was dull: children often expressed their solution by actions and gestures, and didn't share their knowledge with their peers.

Another related study was conducted by Elliott (1993), who examined the impact of metacognitive scaffolding on mathematics thinking of higher and lower achieving kindergarten children. In this study, the metacognitive group was compared with a control group whose teachers were instructed to use their "best practice" and to seek guidance from the curriculum and resource books. Elliott (1993) further explains that, "typically, 'best practice' involves direct guidance with minimal teacher involvement other than that of encouraging, managerial or confirmatory nature. In contrast to the metacognitive approach, there was little modelling of relevant process, little if any discussion of 'why' or 'how', little focus on planning, monitoring, and evaluation, and little emphasis on peer interaction." Elliott reported that children who participated in metacognitive-guided mathematics sessions scored significantly higher on mathematics achievement test compared to the control group. Of particular interest is the positive effect of the metacognitive approach on the lower achieving children.

Primary and secondary school students

A large amount of information is now available on the impact of IMPROVE and other metacognitive interventions on the mathematics achievement of primary and secondary school students. Most of these studies test achievement in arithmetic and algebra, but there are also a small number of studies on achievement in geometry. Positive effects were evident for routine problems as well as for authentic and CUN tasks (e.g. Stillman and Mevarech, 2010).

Achievement in arithmetic and algebra

Many of the studies on the effects of metacognitive interventions indicate that students in elementary schools who studied mathematics via IMPROVE were better able to solve basic, as well as complex problems, and transfer their knowledge to new tasks (Mevarech et al., 2010; Kramarski et al., 2010). In secondary schools the findings were quite similar: IMPROVE students outperformed the control group on various mathematics tasks, including routine and non-routine problems (e.g. Mevarech, 1999), mathematics modelling, translating authentic real-life situations into mathematical expressions, and the finding of mathematics patterns and generalisations (Mevarech, Tabuk and Sinai, 2006). In some of these studies the impact of IMPROVE was found not only on cognitive and metacognitive performance, but also on mathematics anxiety (Kramarski et al., 2010), motivation (Kramarski, 2011), or self-efficacy (Kramarski, 2008) (see Chapter 6 on the use of metacognitive approaches to foster social and emotional skills). Figure 5.1 provides an example of the effects of IMPROVE compared to a control group on third graders' performance on different kinds of mathematics tasks (Mevarech et al., 2010). Figure 5.2 shows the impact on early secondary school students' mathematics reasoning (Mevarech and Kramarski, 1997). In both examples, although no significant differences were found between the groups prior to the beginning of the study, significant differences were found between the groups after students were exposed to IMPROVE. Furthermore, lower and middle achievers benefitted from IMPROVE, but not at the expense of the higher achievers.

Figure 5.1. **Impact of IMPROVE on third graders' mathematics achievement**

Mean scores by condition

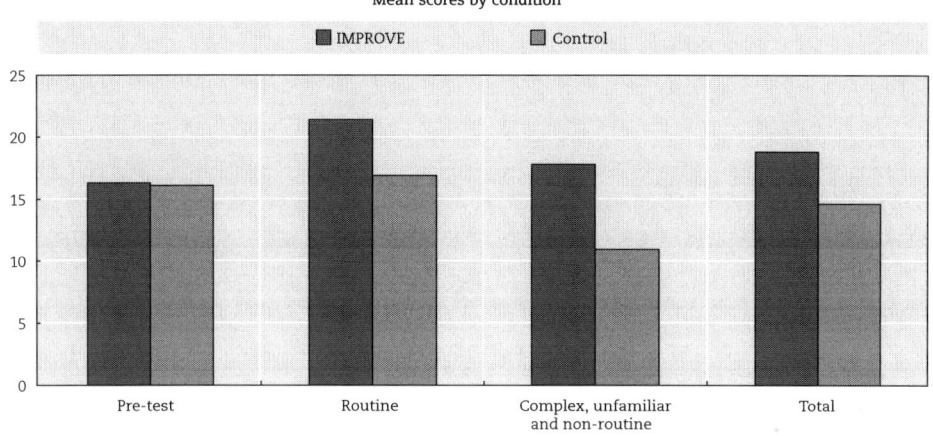

StatLink http://dx.doi.org/10.1787/888933148903

Source: Mevarech, Terkieltaub, Vinberger & Nevet (2010), "The effects of meta-cognitive instruction on third and sixth graders solving word problems", *ZDM International Journal on Mathematics Education,* Vol. 42(2), pp. 195-203.

Figure 5.2. **Impact of IMPROVE on the mathematical reasoning of early lower secondary students**

Mathematical reasoning of higher and lower achievers by conditions

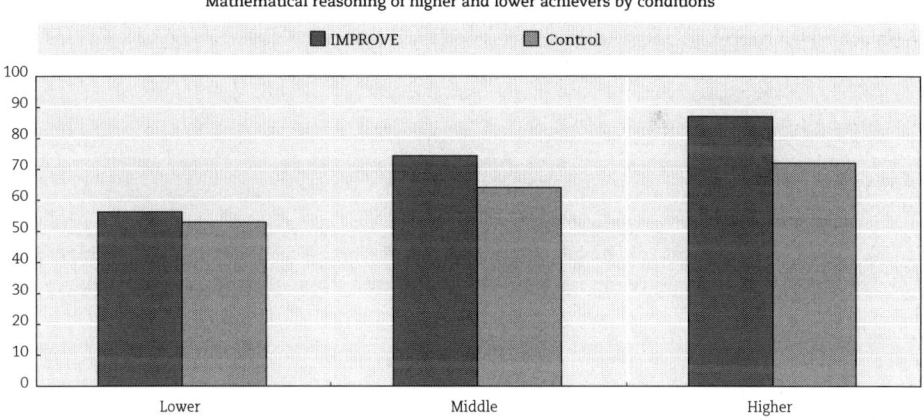

StatLink http://dx.doi.org/10.1787/888933148917

Source: Mevarech and Kramarski (1997), "IMPROVE: A multidimensional method for teaching mathematics in heterogeneous classrooms", *American Educational Research Journal,* Vol. 34(2), pp. 365-395.

Interestingly, the findings indicate that the effects of IMPROVE were stronger on the more complex tasks, whereas the differences between the IMPROVE groups and control groups on routine, "typical" textbook problems were either insignificant or relatively small (Kramarski et al., 2010; Mevarech et al., 2010). The same phenomenon was found among secondary school students (Kramarski, 2011) and college students

(Mevarech and Fridkin, 2006): at all levels, the effects of the metacognitive intervention were stronger on the more complex tasks compared with routine problems. This is not surprising. Quite frequently, students solve the routine textbook problems automatically; they do not have to plan in advance or to monitor and control their solution processes. In contrast, CUN problems cannot be solved without activating metacognitive processes. Consequently, explicit training in how, when, and why one has to apply metacognitive strategies is crucial for these types of problems. The positive effects of IMPROVE on routine and CUN problem solving are in line with the results of Cohors-Fresenborg et al. (2010) who focused on tenth graders and university students.

Several studies examined the effects of other metacognitive pedagogical models. For example, Panaoura, Demetriou and Gagatsis (2010) evaluated the impact of the metacognitive instruction based on the model of Verschaffel, Greer and DeCorte (2000) described in Chapter 4. They found that the metacognitive intervention enhanced fifth graders' self-regulatory strategies and their mathematical performance. Panaoura et al. concluded that the use of an explicit metacognitive model created a powerful learning environment in which students were inspired by their own positive experiences.

Adibnia and Putt (1998) examined the effects of a metacognitive intervention rooted in the Garofalo and Lester model (1985) that includes four steps similar to those described by Schoenfeld (1985): orientation (understanding the problem), organisation (planning and choosing actions), execution (regulating behaviour to conform plans) and verification (evaluating decisions and outcomes). Adibnia and Putt (1998) reported greater improvement in sixth graders' mathematics achievement of the experimental group compared to the control group. Furthermore, higher-ability students appeared to gain more from the experimental instruction than lower-ability students.

Pennequin, Sorel, Nanty and Fontaine (2010) compared the effects of metacognitive training in accordance with Schraw's model (Schraw, 1998) to that of a control group on the students' ability to solve word problems. The results indicated that students in the training group had significantly higher post-test metacognitive knowledge, metacognitive skills and maths problem-solving scores. In addition, the metacognitive training was particularly beneficial for the lower achievers. The training enabled the lower achievers to make progress and solve the same number of problems in the post-test after the training as the typical children solved on the pre-training test.

Many other researchers applied metacognitive questions as a means to enhance the activation of metacognition. For example, Cardelle-Elawar (1995, p. 85) encouraged teachers to ask metacognitive questions such as:

- Do I understand the meaning of the words in this problem? What is the question?

- Do I have all the information needed to solve the problem? What type of information do I need?

- Do I know how to organise the information to solve the problem? Which steps should I take? What do I do first?

- How should I calculate the solution? With which operations do I have difficulty?

The study found that "these questions prompted teachers to focus on the specific steps needed to solve a problem by developing a discourse intended to increase their awareness of potential difficulties that their students might encounter during the process of solving the problem" (Cardelle-Elaware, 1995, pp. 85-86). Indeed, Cardelle-Elaware (1995) found that using this series of metacognitive questions in regular classrooms with a majority of lower achievers significantly enhanced mathematics achievement of the experimental group, independent of grade levels.

The large number of studies that focused on metacognitive interventions led researchers (e.g. Hattie, 1992; Dignath and Buettner, 2008; Dignath et al., 2008) to evaluate the overall effects on mathematics achievement by using meta-analysis techniques – statistical methods that focus on contrasting and combining results from a large number of studies (experimental and quasi-experimental) in order to identify the mean effect size of all the studies that were examined. In meta-analysis, each particular study compares the mean score of the experimental groups to that of the control groups, and a total mean effect size (ES) is calculated (for more information about meta-analysis see *http://www.learningandteaching.info/teaching/what_works.htm*).

Hattie (1992), Dignath and Buettner (2008), and Dignath et al. (2008) reported positive effect sizes of the metacognitive pedagogies on schooling outcomes, indicating that overall the experimental groups that were exposed to metacognitive intervention significantly outperformed the control group.

Dignath and Buettner (2008) and Dignath et al. (2008) went one step further by focusing on the effects of these programs on mathematic achievement. Inspired by the new standards regarding the fostering of life-long learning (EU Council, 2002), they conducted a meta-analysis study that calculated the effect sizes of various self-regulated learning (SRL) pedagogical methods that aim at enhancing cognition, metacognition (planning, monitoring, and evaluating personal progress), and motivation to learn (e.g. Pintrich, 2000; Zimmerman, 2000). Among the various methods, Dingnath and Buettner analysed those that provided metacognitive training. They believe that "fostering self-regulated learning among students would not only improve schooling outcomes, but throughout their entire working life" (2008, p. 232). The authors distinguish between school disciplines (reading, maths, etc.), grade levels (primary versus secondary schools), and whether the intervention was implemented by the classroom teacher or the researcher.

To calculate the effect sizes of these methods, Dignath and Butner (2008) synthesised 74 studies. Of these, 49 studies were implemented in primary schools (first to sixth grades) and 35 studies in secondary schools (seventh to twelfth grades); altogether they calculated 357 effect sizes. Following Schraw (1998), they defined the instruction of metacognitive strategies as including three types of metacognitive strategies: planning, monitoring and evaluation. They added a separate category, metacognitive reflection, namely understanding how to use a strategy, the conditions under which the strategy is most useful and the benefits of using it.

Their findings (Dignath and Buettner, 2008; Dignath et al., 2008) are fascinating: metacognitive interventions attained higher effect sizes in primary schools than in secondary schools (with an effect size for overall academic performance of .61 and .54 standard deviations, respectively), and effect sizes were also higher in primary school mathematics interventions than in reading and writing, and other subject areas (effect sizes for primary schools were .96, .44 and .64 standard deviations for mathematics, reading and writing, and other subjects respectively). In contrast, the effect sizes in mathematics and science were lower in secondary schools compared with reading-writing, and other subjects (effect sizes for secondary schools were .23, .92 and .050 standard deviations, respectively). These findings agree with another meta-analysis study conducted ten years earlier (Hattie et al., 1996) that also showed stronger effects of self-regulated learning intervention on the general academic skills of primary school students compared with secondary school ones.

Why was the effect size greater for primary school students compared with secondary school level, especially in the area of mathematics? There are at least two possible reasons for this finding. First, younger students are more flexible and open to change than older students. Second, younger children might be more in need of such instruction because they lack metacognitive strategies, whereas older students may have automated many of the metacognitive strategies needed for solving maths problems (Veenman et al., 2006).

These empirical findings indicate that primary school children can and do engage in metacognitive activities to self-regulate their learning in general, and mathematics learning in particular (Dignath et al., 2008; Perry et al., 2004; Perry, VandeKamp, Mercer and Nordby, 2002). Hattie et al. (1996) concluded that most of the advantage of metacognitive training is gained at the beginning of children's schooling because during these first crucial years students set up learning strategies and self-efficacy attitudes which are easier to change than when students have already developed disadvantageous learning styles and learning behaviour. This is not to say that secondary school students do not need metacognitive guidance; the smaller effect size for secondary school simply highlights the differences between the two age groups.

Achievement in geometry

Line, shapes, and objects are found everywhere: houses, bridges, global positioning systems (GPS) data, maps, city plans, crystals, snowflakes, etc. They can be static or dynamic, represented as images, real objects, or models. No wonder that "shapes and space" are sometimes identified as one of the four big ideas in mathematics (the other three are: quantity, change and relationships, and uncertainty) (OECD/ UNESCO Institute for Statistics, 2003). The study of geometry is, thus, fundamental:

> Students should recognise shapes in different representations and different dimensions Students must be able to understand relative positions of objects and to be aware of how they see things and why they see them this way. Students must learn to navigate through space and through constructions

and shapes. Students should be able to understand the relation between shapes and images or visual representations... They must also understand how three-dimensional objects can be represented in two dimensions, how shadows are formed and interpreted and what "perspective" is and how it functions. (OECD, 2007, p.24).

According to the framework of the OECD Programme for International Student Assessment (PISA), in the study of shapes and constructions students should plan ahead, look for similarities and differences as they analyse the components of the forms, find strategies of representations, monitor and control their solutions, and reflect on the inputs, processes, and outcomes (OECD, 2007). Thus, the use of metacognitive processes seems to be necessary in studying geometry, rather than just "nice to have".

However, while there are plenty of studies that have explored the effects of metacognitive pedagogies on students' abilities to solve word problems, the number of studies that have focused on geometry is rather small. This is quite surprising because: 1) geometry is an integral part of the mathematics curriculum from kindergarten up to the end of secondary school; 2) geometry is considered to be one of the more difficult subjects among the various mathematics areas (TIMSS, 1997) probably because it requires rigorous proofs based on formal mathematics language (at least in high schools), exact definitions of shapes and objects, generalisations, and abstract reasoning; and 3) metacognitive pedagogies have proven to be effective in enhancing students' problem solving, particularly CUN problems.

Generally, the few studies that have looked at the interface between geometry and metacognition could be classified into two categories: one examining the metacognitive skills that are activated in solving geometry problems, and the other exploring the effects of metacognitive pedagogies on students' achievement in geometry. Regarding the first category, Lucangeli and Cornoldi (1997) explored the relationships between metacognitive monitoring processes and problem solving in various mathematics areas. Assessing third and fourth graders (a sample size of 397 and 394 students, respectively) using standardised mathematics tests they found that numerical and geometrical problem-solving abilities were most strongly related to metacognitive capabilities, particularly to awareness of the monitoring and control processes during the test execution (Lucangeli and Cornoldi, 1997).

About a decade later, Yang (2012) explored the structural relationship between students' use of metacognitive reading strategies and their reading comprehension of geometry proofs. Assessing a sample of 533 ninth graders, Yang found that the use of metacognitive reading strategies was related to students' reading comprehension of geometry proofs. As expected, "good comprehenders tended to employ more metacognitive reading strategies for planning and monitoring comprehension and more cognitive reading strategies for elaborating proof compared with the moderate comprehenders, who in turn employed these strategies more often compared with the poor comprehenders" (Yang, 2012, p. 307).

Using a computer-supported collaborative problem-solving environment, Hurme, Palonen, and Järvelä (2006) analysed student interactions while solving tasks involving polygons in a geometry course. Interestingly, while metacognitive activities varied among participants, the researchers never encountered some aspects of metacognition, such as planning. This might be due to the fact that students were not instructed how to activate metacognitive processes while solving the geometrical problems.

Three studies (Hauptman, 2010; Schwonk et al., 2013; Mevarech et al., 2013) may illustrate the effects of metacognitive pedagogies on student achievement in geometry. While all three studies used some version of metacognitive self-addressed questioning as described by IMPROVE, they varied in terms of the educational environments (virtual reality, Cognitive Tutor, or no computer support, respectively), as well as the geometry content (three dimensions, angles and intersection lines, and trapezoids, respectively).

Being aware of the difficulties students face in studying space geometry, Hauptman (2010) developed a software environment based on a virtual reality (VR) technique that enables the user to build spatial images and manipulate them. The research also explored the additional effects of training students to use self-regulated questioning (SRQ) as described by IMPROVE. The participants in this study were 192 tenth grade Israeli students who were randomly assigned into four groups: being exposed to VR + SRQ, VR with no SRQ, SRQ with no VR, and a control group with no VR and no SRQ. The tested outcomes were mental rotation measured using the Mental Rotation Tests (MRT) and spatial-visual reasoning using the Aptitude Profile of Spatial-Visual Reasoning Test (APTS-E). While no significant differences were found between the four groups prior to the beginning of the study, the students who were exposed to VR and SRQ outperformed the VR with no SRQ group who in turn outperformed the other two groups on the twotests. Hauptman concludes that the self-regulated questioning enhances geometry reasoning whether it is embedded within a virtual reality environment or in a traditional classroom. Figures 5.3 and 5.4 present the mean scores of the four groups on the MRT and APTS-E tests.

A recent study by Schwonke, Ertelt, Otieno, Renkle, Aleven and Salden (2013) explored the effects of metacognitive knowledge on students' achievements in geometry course. The participants were 60 German Realschule eighth graders who studied the Angles and Lines unit with the aid of some software called Geometry Cognitive Tutor. Students studied either with or without metacognitive support (30 students in each group). The metacognitive support included a set of six hints arranged in two groups: 1) how do I solve the problem (e.g. "what are the known values in the problem text? Can you locate them in the geometry diagram?"); and 2) what do I do when I get stuck? (e.g. "when you need to find out about the relevant mathematical principle consult the glossary tool"). The results indicated that the metacognitive group outperformed the other group on geometry achievement tests including conceptual and procedural knowledge. Furthermore, the metacognitive support made learning more efficient; students could have less learning time without impairing outcomes. Finally, students with low prior knowledge exposed to the metacognitive support developed deeper conceptual understanding than the other group. The authors concluded that "a lack of metacognitive conditional knowledge

Figure 5.3. **Impact of virtual reality and self-regulated questioning on mental rotation abilities**

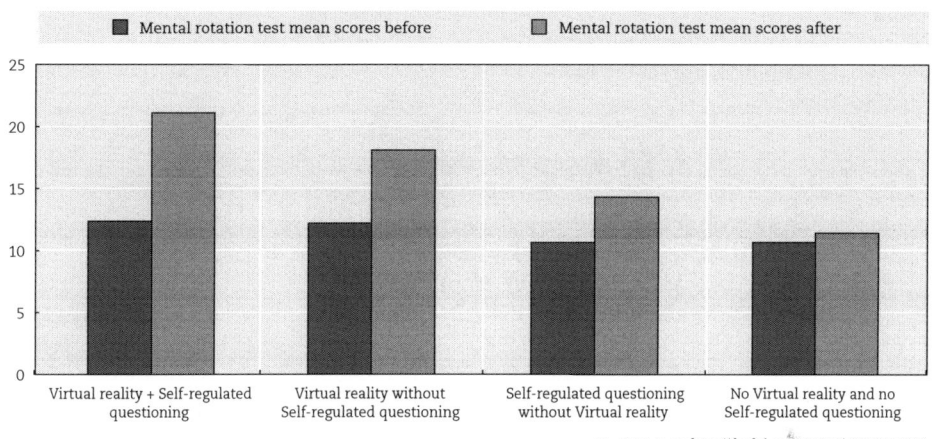

StatLink ᵃⁱᵖ http://dx.doi.org/10.1787/888933148926

Source: Hauptman (2010), "Enhancement of spatial thinking with Virtual Spaces 1.0", *Computers and Education*, Vol. 54(1), pp. 123-135.

Figure 5.4. **Impact of virtual reality and self-regulated questioning on spatial-visual reasoning**

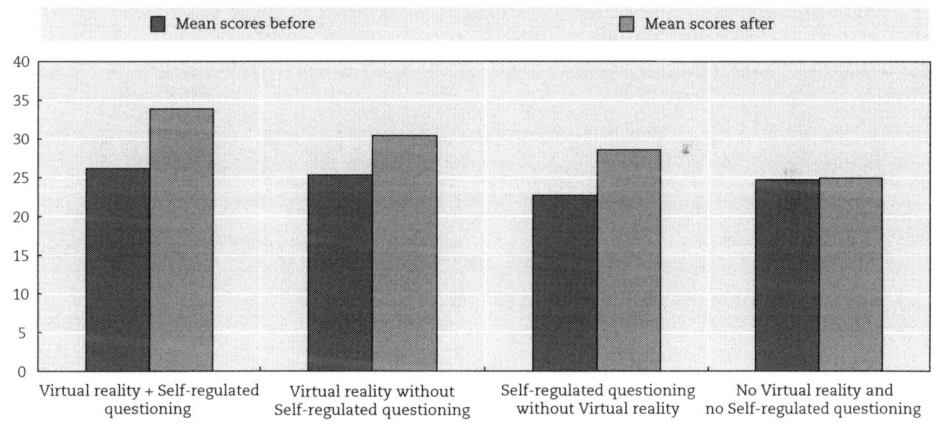

StatLink ᵃⁱᵖ http://dx.doi.org/10.1787/888933148934

Source: Hauptman (2010), "Enhancement of spatial thinking with Virtual Spaces 1.0", *Computers and Education*, Vol. 54(1), pp. 123-135.

(i.e. in which situation to use which help facility) can account for learning difficulty in computer-based learning environments" (Schwonke et al., 2013, p. 136).

In contrast to the previous two studies, Mevarech, Gold, Gitelman and Gal-Fogel (2013) examined the immediate and lasting effects of metacognitive instruction implemented via IMPROVE on students' judgment of learning (JOL) and its accuracy as assessed by an achievement test in geometry (see Chapter 8 for more information on judgement of learning). According to the cue theory, individuals are more likely to judge their ability to remember an item if they are familiar with the item (Koriat, 2008). Based on this theory, the researchers hypothesised that IMPROVE would enhance students' understanding which in turn would facilitate their JOL and its accuracy. The participants were 90 ninth grade Israeli students (four classrooms). Whole classrooms were randomly assigned into one of two conditions: studying with or without IMPROVE (N=48 and 42, respectively). The instructional unit was trapezoids, including definitions, proofs and computations of angles, perimeters and areas. Measurements included achievement tests in geometry (pre- and post-tests), a JOL questionnaire (pre- and post-assessments) and observations. The results indicated that although the IMPROVE group scored significantly lower than the control group on the present, at the end of the study the IMPROVE group outperformed the control group on the achievement tests (Mean = 73.4 and 50.6, respectively; Standard Deviation = 23.2 and 27.2, respectively; $F(1,87) = 22.70$, $p<.001$, after controlling for prior achievement). In addition, at the end of the study, the JOL ratings of the experimental group were significantly higher than that of the control group (Mean = 3.66 and 2.91, respectively; Standard Deviation = .71 and .90, respectively) controlling for prior ratings $F(1,87) = 17.56$, $p<.001$).

In summary, students learning geometry who are more able to monitor, control and regulate their learning are better able to solve the given tasks, just as was found in the areas of arithmetic and algebra. Interestingly, these positive effects were evident whether students were exposed to metacognitive pedagogies in a virtual reality environment, through Cognitive Tutor, or traditional instruction with no computer support, they outperformed their counterparts who were given no metacognitive support.

Research on the impact of metacognitive pedagogies on student achievement in geometry is only at its beginning and many issues are still open. For example, none of the studies reviewed above distinguished between CUN and routine tasks in geometry. Furthermore, the teaching of geometry, even more than the teaching of other areas in mathematics, is based on "doing" (e.g. using manipulations such as blocks, constructing shapes and objects, and of course, planning and carrying out the plans). It is quite possible that metacognitive pedagogies need to be modified to meet the specific purposes of teaching geometry at different grade levels. Finally, none of these studies addressed the differential effects of the metacognitive pedagogies on gender. Since boys have more advanced spatial skills than girls especially during adolescence (Leahey and Guo, 2001), it would be interesting to explore the extent to which metacognitive pedagogies could decrease the gender gap. All these issues merit future research.

Impact on the solution of authentic tasks

Authentic tasks are a specific type of CUN problem. Authentic tasks employ realistic data, provide rich information about the situation described, can be approached in different ways, and often use different representations. Mueller (2012) considers authentic mathematics tasks as those which ask students to apply standard-driven knowledge and skills to real-world challenges. Obviously, what might be considered as authentic or CUN task for one, might turned out to be routine and familiar for another. Therefore, according to Muller (2012) the attributes of traditional and authentic tasks lie in a continuum varying from:

- selecting responses.............. to .... performing a task

- contrived .... to .... real life

- recall/recognition .... to .... construction/ application

- teacher structured .... to .... student structured

- indirect evidence.................. to .... direct evidence

The NCTM (2000), PISA (OECD, 2003, 2012), as well as many other institutions or programmes, emphasise again and again the importance of training students to solve authentic tasks. The reason is threefold: 1) to promote mathematics content and procedural knowledge; 2) to prepare students to apply mathematics in real-life contexts; and 3) to increase students' motivation by familiarising them with real-life uses of mathematics.

Many students, both lower and higher achievers, face difficulties in solving authentic tasks (OECD, 2003). They raise difficulties at all stages of the solution process, from the very first stage of understanding what the problem is all about, through planning the solution process and selecting appropriate strategies, to reflecting on the solution and deciding whether the solution makes sense (Verschaffel et al., 2000).

Some students, particularly lower achievers, do not see the task as a whole and thus focus only on parts of the task (e.g. Lester, 1994). According to Cardelle-Elawar (1995) and Frye (1987), lower achievers rapidly read the task at the expense of fully comprehending it. They do not recognise that there might be more than one correct way to solve the task, and they are uncertain about how to calculate and verify the solution. Verschaffel et al. (2000) indicate that lower achievers have difficulties in reorganising the given information and distinguishing between relevant and irrelevant information.

Higher achievers also face difficulties in solving authentic tasks, although the difficulties are different. Higher achievers give up easily because ready-made algorithms are not available for solving the authentic task, and they also have difficulties in transferring what they know about standard tasks to the novel, authentic tasks (Frye, 1987; Verschaffel et al., 2000).

Difficulties in solving authentic tasks were also identified in pre-service mathematics teachers (Yimer and Ellerton, 2010). Yimer and Ellerton showed there were significant differences in participants' cognitive and metacognitive processes in solving the authentic problems, and also individual solvers showed significant differences in approaching different tasks. On the basis of these analyses, Yimer and Ellerton proposed a five-phase model for identifying the metacognitive and cognitive processes involved in solving authentic problems, and the apparent difficulties in each phase: engagement, transformation-formulation, implementation, evaluation and internalisation.

Since many of the difficulties associated with solving authentic tasks lie in students' inability to control, monitor and reflect on their solution processes, Kramarski, Mevarech and Arami (2002) examined the extent to which IMPROVE has the potential to facilitate the solution of authentic tasks. In this study, whole seventh grade classrooms were randomly assigned into two groups who studied in co-operative settings. One group was exposed to metacognitive scaffolding provided by IMPROVE, and the other studied "traditionally" with no metacognitive intervention. At the end of the semester, all students were administered the Pizza task (see Box 1.1).

The findings indicate that IMPROVE students outperformed the other group on both routine problems and on the authentic task. The positive effects of IMPROVE were observed on four criteria: 1) referring to all of the data; 2) organising information; 3) processing information; and 4) making decision (i.e. solving the problem) and justifying it. For example, qualitative analyses of students' answers indicated that IMPROVE students referred to all the information given in the text by offering a combination of different kinds of pizza and carrying out multiple mathematics operations, whereas none of the students in the other group suggested a similar solution. Instead, the later used rather simple strategies based only on multiplication of the numbers. Figures 5.5 and 5.6 present the effects of IMPROVE compared with that of the control group on students' performance on routine and authentic tasks, and on each component of the authentic task's solution.

Figure 5.5. **Impact of IMPROVE on authentic and routine tasks**

Mean scores by condition

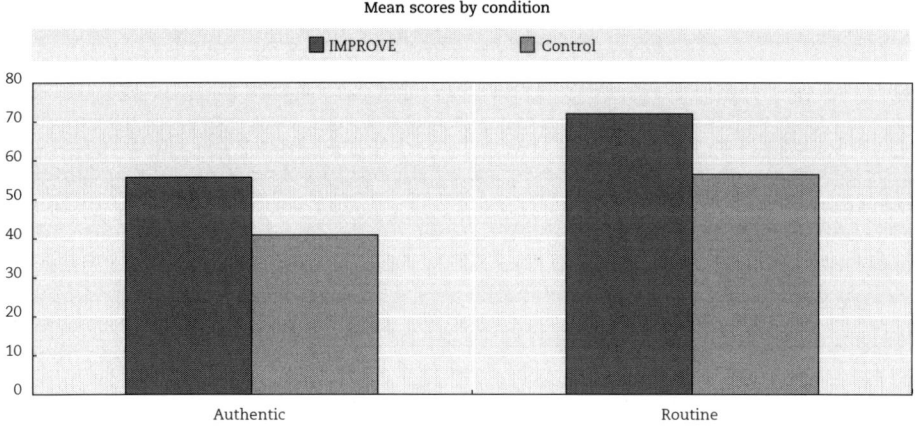

StatLink http://dx.doi.org/10.1787/888933148949

Source: Kramarski, Mevarech and Arami, (2002), "The effects of metacognitive training on solving mathematical authentic tasks", *Educational Studies in Mathematics*, Vol. 49, pp. 225-250.

Figure 5.6. **Impact of IMPROVE on all components of solving an authentic task**

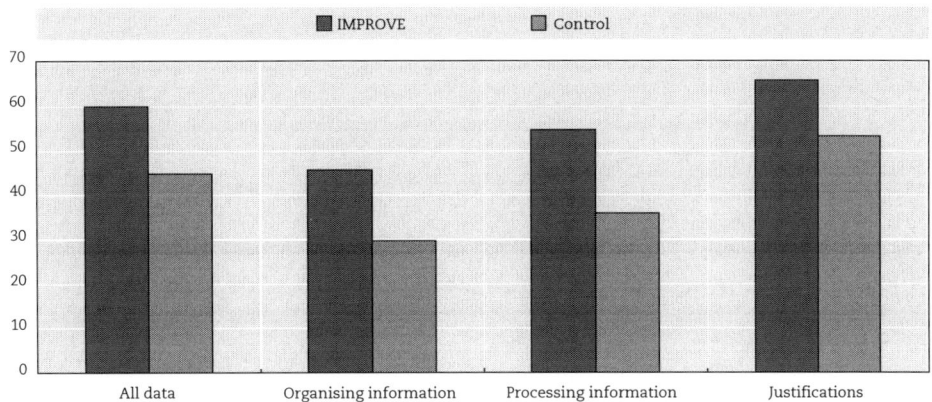

StatLink ⬛📊 http://dx.doi.org/10.1787/888933148952

Source: Kramarski, Mevarech and Arami, (2002), "The effects of metacognitive training on solving mathematical authentic tasks", Educational Studies in Mathematics, Vol. 49, pp. 225-250.

How and why did the meta-cognitive self-directed questioning assist students in solving the authentic task? We can only interpret the findings. By being trained to think what the problem is all about, the comprehension question probably led students to focus on the mathematical construct of the tasks as well as on all the given information by distinguishing between the relevant and irrelevant information. The connection question might familiarise the novel problem by relating it to the problems solved in the past. The strategic question perhaps led students to reorganise the information and represent it in various forms. Finally, the reflection question most likely ties the whole process by leading students to offer various solutions, ask themselves if the offers make sense, and whether the problem can be solved differently.

College students

While there is much disagreement about the extent to which young children can activate metacognitive processes, there is almost a consensus that adults have already acquired the basic components of metacognition. Schraw et al. (2006) indicate that most adults can regulate their learning: they can plan, monitor, control, debug and reflect on their cognitive activities. Adults have also declarative, strategic and conditional knowledge: they know when, how and why to activate problem solving strategies. It is believed that school as well as natural development assists in training people to apply cognitive and metacognitive processes in solving problems.

Yet, although adults are likely to have acquired metacognitive skills, recent studies have indicated that this is not always the case. For example, McCabe (2011) showed that undergraduates are largely unaware of several specific strategies that could assist them in recalling the course information and that training has the potential to improve metacognitive judgments in these domains. The need to train students to implement metacognitive processes has been evident in different disciplines, including mathematics, reading and medicine (Lajoie et al., 2013).

To address this issue, Mevarech and Fridkin (2006) conducted a study in which college students who took a course in mathematics were randomly assigned into one of two groups taught by the same instructor: one group was trained via IMPROVE, and the other studied in a traditional way with no metacognitive intervention. Three measures were used in this study: achievement test that assessed maths knowledge and reasoning, metacognitive awareness inventory (MAI) designed by Schraw and Davidson (1994) (see Chapter 2), and a maths metacognitive questionnaires. The findings indicate that IMPROVE students outperformed the control group on mathematics achievement and mathematics reasoning. In addition, the IMPROVE students reported applying higher levels of metacognitive processes in solving mathematics (i.e. domain-specific metacognition) as well as non-mathematics problems (i.e. general metacognition). Figure 5.7 presents the college students' mathematics achievement by learning conditions, and Figures 5.8 and 5.9 show the mean scores of these students on the two main metacognitive components (knowledge of cognition and regulation of cognition, respectively) by learning conditions.

Figure 5.7. **Impact of IMPROVE on college students' mathematics achievement**

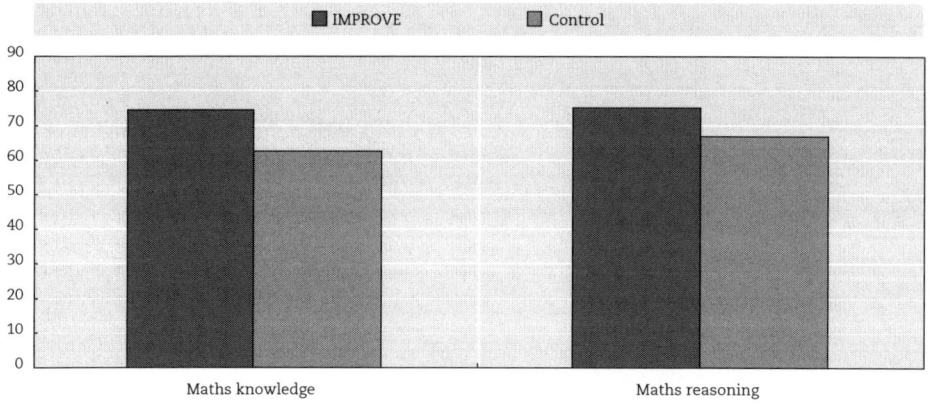

 StatLink http://dx.doi.org/10.1787/888933148960

Source: Mevarech and Fridkin (2006), "The effects of IMPROVE on mathematical knowledge, mathematical reasoning and meta-cognition", *Metacognition Learning*, Vol. 1, pp. 85-97.

Figure 5.8. **Impact of IMPROVE on college students' knowledge of cognition**

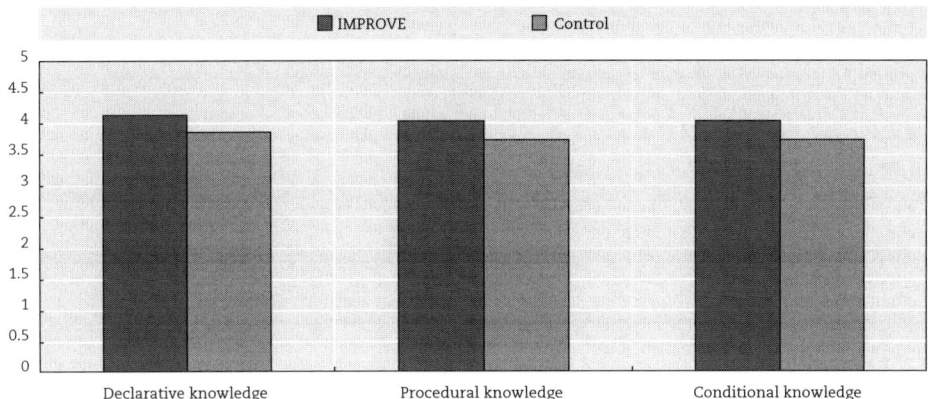

StatLink http://dx.doi.org/10.1787/888933148979

Source: Mevarech and Fridkin (2006), "The effects of IMPROVE on mathematical knowledge, mathematical reasoning and meta-cognition", *Metacognition Learning*, Vol. 1, pp. 85-97.

Figure 5.9. **Impact of IMPROVE on college students' regulation of cognition**

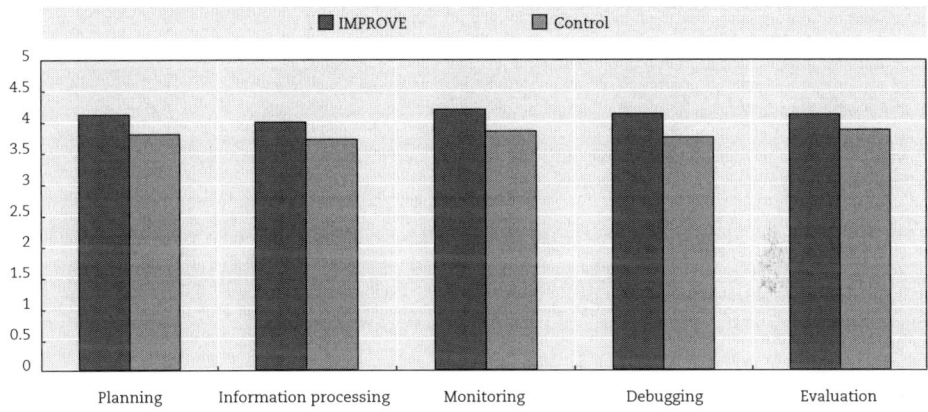

StatLink http://dx.doi.org/10.1787/888933148986

Source: Mevarech and Fridkin (2006), "The effects of IMPROVE on mathematical knowledge, mathematical reasoning and meta-cognition", *Metacognition Learning*, Vol. 1, pp. 85-97.

These findings are in line with other studies (e.g. Lovett, 2008; Subocz, 2007; Davis, 2009). Lovett (2008) indicates that metacognitive instruction based on self monitoring among college students improves performance even after a delay, encourages the use of metacognitive strategies, and increases self-confidence and attitudes towards mathematics. This supports the conclusion that effective learning involves planning and goal-setting, monitoring and adapting one's progress (e.g. Winne, 1995). Several Ph.D. dissertations (e.g. Subocz, 2007; Davis, 2009) reported similar findings: being exposed to metacognitive strategy interventions in community colleges decreased students' failure rate, improved their attitudes towards mathematics, and increased their use of higher-level metacognitive skills in solving word problems compared to solving equations without any context.

The examples described above show similar findings to those indicated in a meta-analysis study conducted by Ragosta (2010). The study aimed to determine the effectiveness of interventions designed to help college students to acquire self-regulated learning strategies in mathematics courses. The meta-analysis was based on 55 primary studies with a total sample of 6 669 students. The overall effect size for all studies was .335 of a standard deviation, showing that even college students can improve their self-regulated learning.

The impact in high-stakes situations

While there is a lot of evidence showing how students apply metacognitive processes in situations similar to the ones in which they were trained to do so, very little is known at present on the extent to which students use their metacognitive knowledge in novel situations. One may argue that the effects of metacognitive instruction are limited to "here-and-now", and thus "metacognitive processes would not be transferred to situations different from those in which students were trained, particularly not in situations which are highly demanding, stressful, and time constrained" (Mevarech and Amrany, 2009, p. 148). An alternative hypothesis is that students would recognise the added value of metacognitive processes and thus would apply these processes in all situations, including those that are characterised as imposing a high cognitive load.

One highly demanding situation is that in which students are being examined on the matriculation or national exams administered in many countries at the end of secondary school. Teachers and students devote a lot of efforts in preparation for these kinds of exams, and generally do not like to change the well-known traditional instructional method. Examining the effects of IMPROVE at the end of secondary school has practical implications because it underlines the effectiveness of the method in extreme, very demanding, situations.

Mevarech and Amrany (2008) conducted a study in which they used quantitative and qualitative analyses to assess the effects of IMPROVE on mathematics achievement and metacognition of students who took the matriculation exam in mathematics (middle level), compared with a non-treatment control group. Three kinds of measurements were used in this study: mathematics achievement tests based on the matriculation exams for middle level, a metacognitive awareness questionnaire adopted from Schraw and Dennison (1994) and interviews. The mathematics tests and questionnaire were administered twice: prior to the beginning of the study and at the end of the study. The metacognitive awareness questionnaire assessed the two metacognitive components: knowledge of cognition, and regulation of cognition (see Chapter 2). The interviews took place immediately after students completed the matriculation exam, about two months after the end of the intervention. When students came out of the matriculation exam, one of the authors presented the students with problems that were on the exam and asked them to think aloud while solving the problems as they did on the exam. The duration of the study was one semester.

The results indicated that, as with previous studies, the IMPROVE students significantly outperformed the control group on mathematics achievement (Figure 5.10). With regards to the metacognitive awareness questionnaire, the IMPROVE students scored significantly higher on the regulation-of-cognition component (e.g. "I consider several alternatives to a problem before I answer"), but no significant differences were found between the two groups on the knowledge-of-cognition component (e.g. "I understand my intellectual strengths and weakness"). The latter finding might be an indication that by the end of high school many students have already acquired knowledge of cognition: they have theoretical knowledge about effective learning strategies, but they lack training on how to regulate their learning. Analysing the interviews indicated that IMPROVE students applied metacognitive processes in situations beyond those in which they were trained to do so. Furthermore, the positive effects were evident a couple of months after the end of the intervention. However, because different teachers taught the experimental and control groups, we could not rule out teacher effects. Future research may continue studying the effects of metacognitive pedagogies in high-stakes situations.

Figure 5.10. Impact of IMPROVE on secondary school students' high-stakes mathematics achievement

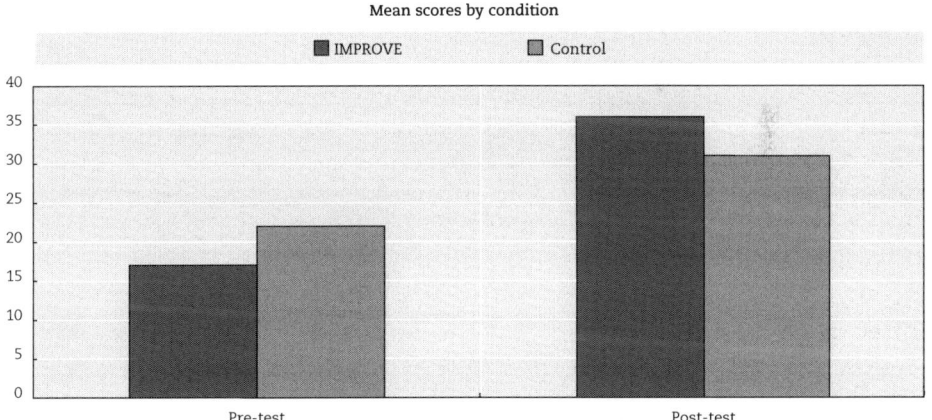

Mean scores by condition

■ IMPROVE　　　■ Control

StatLink ▮▰▰ http://dx.doi.org/10.1787/888933148999

Source: Mevarech and Amrany (2008), "Immediate and delayed effects of meta-cognitive instruction on regulation of cognition and mathematics achievement", *Metacognition Learning*, Vol. 3, pp. 147-157.

To sum up, metacognitive pedagogies have proven to be effective methods to enhance students' mathematics achievement, both in routine and CUN problem solving, and applying metacognitive processes. The effects are found at all grade levels: kindergarten, primary and secondary school, and at university. All the studies reviewed above were implemented in "regular" classrooms, showing the high ecological validity of the metacognitive pedagogies. However, all of these studies with the exception of Mevarech and Amrany (2008), assessed the effects immediately after the end of the intervention. Would similar effects be found on a delayed assessment, or when the metacognitive pedagogy is implemented over a full academic year, or under specific conditions?

Immediate, delayed and lasting effects

A major challenge for the development of certain pedagogies is the transition from the experimental stage into classroom conditions. Educators, policy makers, and researchers are often interested in large-scale studies that continue over a full academic year, and their lasting effects. The effects of such studies are reported below.

Effects over a full academic year

Although the reported effects of IMPROVE and other metacognitive interventions are impressive, the evidence is based on relatively short interventions of about one to three months. Many teachers do not like to change instructional methods during the school year. It is also questionable whether these positive effects resulted from some kind of "Hawthorne effect" related to the initial excitement going with any intervention, and thus would not necessarily be evident when the innovative method is implemented during a full academic year.

In 1997, Mevarech and Kramarski reported the first research on the effects of IMPROVE on seventh graders' mathematics achievement and reasoning. The participants were all ten schools in one city in the centre of Israel. Six of the ten schools were randomly selected to implement IMPROVE and four served as a non-treatment control group. Given the complexity of the study, involving schools, classrooms within schools, and students within classrooms, the data were analysed using a multilevel hierarchical linear model. The study focused on students' mathematical reasoning and their competencies to solve "typical" textbook problems. The following are two examples of reasoning problems administered in this study.

Ron argues that X/X (X not equal 0) always equals 1. Sarah argues that the value of X/X depends on the value of X. Who is correct? Please explain your reasoning.

If $a > 0$ and $b < 0$, is their difference a positive or negative number? Please explain your reasoning.

While no significant differences were found between the two groups prior to the beginning of the study, at the end of the first semester, the IMPROVE group significantly outperformed the control group by almost seven points on the achievement test, and about ten points (a full score) on the reasoning part. The differences between the two groups were found for lower, middle and higher achievers. Similar findings were also reported at the end of the academic year, after students were exposed to the method over a full academic year, as shown in Figure 5.11 below.

Figure 5.11. **Impact of IMPROVE over one academic year**

Mean scores by condition

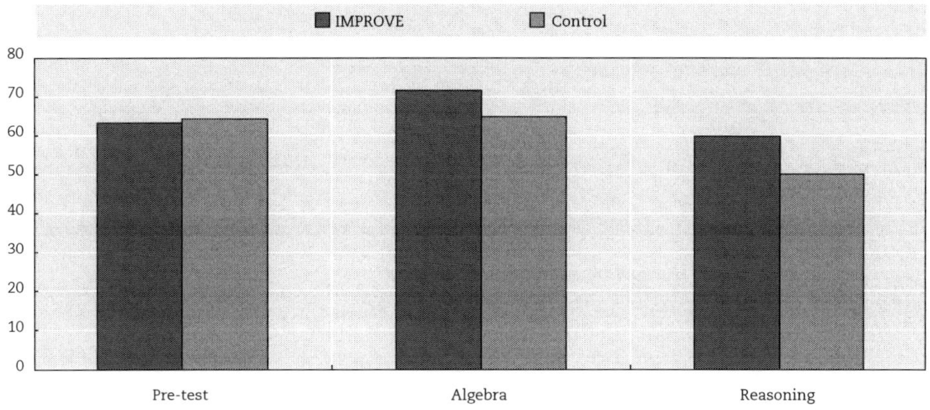

StatLink ⬛🔗 http://dx.doi.org/10.1787/888933149008

Source: Mevarech and Kramarski (1997), "IMPROVE: A multidimensional method for teaching mathematics in heterogeneous classrooms", *American Educational Research Journal*, Vol. 34(2), pp. 365-395.

Immediate, delayed and lasting effects of metacognitive instruction

The issue of "lasting effects" concerns everyone who is involved in innovative teaching methods. Teachers are interested in the extent to which students are capable of recalling what they have learned via the innovative method compared with the control group. Similarly, policy makers who are usually concerned with budgetary issues look for evidence showing the immediate, delayed and lasting effects of a proposed teaching method. Finally, researchers like to know whether the positive effects of the experimental method are still evident long after the experiment has finished. Although the issue of "lasting effects" is important, it is usually difficult to assess, because it requires examining the same students for a relatively long period of time.

Mevarech and Kramarski (2003) analysed the lasting effects of IMPROVE by examining the students a year after they were exposed to the method. Intact eighth grade classrooms were randomly assigned into one of two treatment groups, both implemented in co-operative settings: one group studied algebra via IMPROVE, and the other served as a control group, in which the teacher administered worked-out examples that specified each step of the solution and provided explanations as needed. The worked-out examples were followed by practising problems of the same kind.

According to the school policy, when students are in ninth grade, they are randomly reassigned into new classrooms. Thus, a year later, when participants were in ninth grade, students who were exposed to IMPROVE and those who learned by worked-out examples, studied together in the same classroom.

The ninth grade teachers used the "traditional" teaching method with no metacognitive guidance.

Students' mathematics achievement and reasoning were assessed three times: prior to the beginning of the study that is, in eighth grade; immediately after the students were exposed to IMPROVE; and a year later, during ninth grade. In addition, the students' discourse was videotaped. Data were analysed using quantitative and qualitative methods.

Results indicated no significant differences between the groups on the pre-tests, but significant differences on the immediate and the delayed tests after the intervention. In both cases the IMPROVE students outperformed the worked-out example group. Further analyses showed gains among the IMPROVE students compared with that of the control group on all three discourse criteria: students' verbal explanations of their mathematical reasoning, their algebraic representations of verbal situations, and their algebraic solutions. Moreover, a detailed analysis of each item on the test indicated that the impact of IMPROVE was mainly evident on the more complex problems, whereas no significant differences were found between the groups on the easier tasks. Figure 5.12 presents the students' mean scores for mathematics achievement on before, immediately after and a year after the intervention. It should be noted that the two post-tests (the immediate and delayed) were identical, whereas the pre-tests were different, focusing on general mathematics achievement.

Figure 5.12. **Immediate and lasting impact of IMPROVE on mathematics achievement**

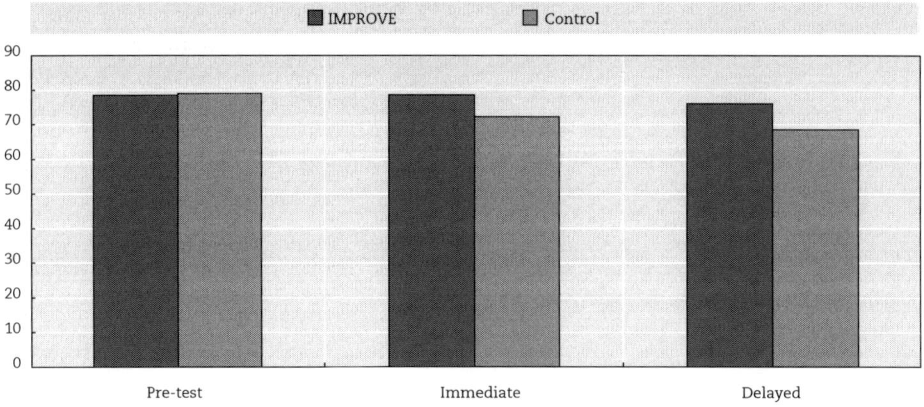

StatLink http://dx.doi.org/10.1787/888933149013

Source: Mevarech and Kramarski (2003), "The effects of metacognitive training versus worked-out examples on students' mathematical reasoning", British Journal of Educational Psychology, Vol. 73(4), pp, 449-471.

The long-lasting effects of a metacognitive intervention on mathematics achievement were also examined by Desoete (2009). In her study, third grade Belgium children were randomly assigned into metacognitive or traditional instruction. The children were assessed in the third and fourth grades on mathematics achievement and metacognition. In this two-year longitudinal study, Desoete showed the advantages of the metacognitive group not just in the third grade but also when the children were in fourth grade: the children in the metacognitive group outperformed the children in the control group on metacognition and mathematics achievement.

While Desoete (2009) and Mevarech and Kramarski (2003) assessed the effects of the metacognitive interventions a year after the intervention ended, Shayer and Adey (2006) conducted an interesting 5-year longitudinal study. They examined the lasting effects of a 2-year metacognitive intervention that was initially implemented when participants were 11 years old, and evaluated the lasting effects by using the British national examinations taken at the age of 16, 3 years after the end of the intervention. Although the intervention was set within the context of science learning, the effects were found in science, mathematics and English. In comparison with control classes the effect sizes were 0.67, 0.72 and 0.69 standard deviations in science, mathematics and English, respectively. Shayer and Adey attributed the big effect sizes to the positive effects of the metacognitive intervention. Clearly, this study is unique not only because of its long duration, but also because it uses national examinations as an assessment tool, rather than teacher or researcher-made tests.

Another longitudinal study was conducted by Weiss and Pasley (2004) who tried to identify "what is high-quality instruction?" They observed 364 representative mathematics and science lessons over 18 months in "natural" settings with no interventions. The authors documented, analysed and assessed lessons according to the following indicators: the quality of the mathematics and science content, the quality of implementation, and the extent to which the classroom culture facilitated learning. The observers rated individual indicators in each area on a scale of one to five, and then looked across these indicators to categorise the lesson's overall quality as low, medium, or high. Weiss and Pasley indicated that one of the most effective components of high-quality instruction (in their words "crucial") relates to teachers' questioning – "the kind that monitors students understanding of new ideas and encourages students to think more deeply" (Weiss and Pasley, 2004, p. 26).

To sum up, IMPROVE and similar metacognitive pedagogies have been successfully implemented over one or two academic years. These methods showed positive effects on students' mathematics achievement, both when assessed immediately after the intervention had ended and after a delay, during which students were no longer exposed to the metacognitive scaffolding. In all these studies, the positive effects of the metacognitive pedagogies were higher than that of the control groups who were not exposed to metacognitive scaffolding.

What conditions work best for metacognitive instructional models?

Are co-operative settings required?

Originally, many of the metacognitive interventions, including IMPROVE, were designed to be implemented in co-operative settings (e.g. Mevarech and Kramarski, 1997; Verschaffel, 1999). The rationale for this is twofold. First, co-operative settings seem to be a natural environment for students to articulate their mathematical reasoning and enhance mathematical communications (NCTM, 2000; Schoenfeld, 1992). Second, as seen in Chapter 3, a large body of research indicates that co-operative learning is an effective environment for enhancing mathematics achievement in general and mathematics reasoning in particular, if students are guided how to work in small groups and apply metacognitive processes (e.g. De Corte, Verschaffel and Eynde, 2000). But is it really necessary to embed metacognitive instruction in co-operative settings? What does each component contribute individually and in combination to students' mathematical achievement and reasoning? To address this question, Kramarski and Mevarech (2003) designed a study in which they compared four conditions. One group studied algebra via metacognitive instruction embedded within co-operative learning. The second group was exposed to metacognitive instruction in individualised settings. The third group studied the same materials in co-operative settings with no metacognitive guidance. And the fourth group studied in individualised settings with no metacognitive guidance. The separation into four treatment groups enabled Kramarski and Mevarech to study the unique contribution of each component on various mathematical skills, including: mathematical creativity and transfer of knowledge.

Twelve eighth grade classrooms (totalling 384 students) were randomly assigned into one of the four groups described above. All groups studied the same topics for the same duration of time, and all used the same textbook. No significant differences were found between them on mathematics achievement prior to the beginning of the study.

The results indicated that students who were exposed to the combined method (metacognitive instruction implemented in a co-operative setting) scored significantly higher on the routine, "typical" achievement test than the group provided with metacognitive guidance in individualised settings, who in turn outperformed the two groups with no metacognitive guidance. No significant differences were found between the latter. The IMPROVE students also outperformed the other groups on measures of creativity (fluency and flexibility) and on the transfer task that assessed students' ability to use their knowledge in novel situations that have not been introduced in class. In all cases, students who were exposed to metacognitive guidance embedded within co-operative settings were more fluent and more flexible compared with the other groups. Figure 5.13 below presents students' learning gain and mean scores on mathematics achievement by time (pre- and post-intervention tests) and learning conditions.

Figure 5.13. **Impact of metacognitive guidance and co-operative learning on mathematics achievement**

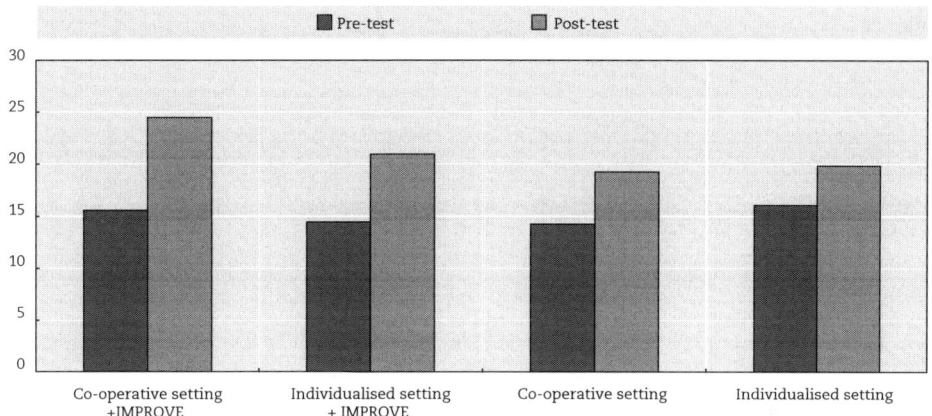

StatLink http://dx.doi.org/10.1787/888933149028

Source: Kramarski and Mevarech (2003), "Enhancing mathematical reasoning in the classroom: Effects of cooperative learning and metacognitive training", *American Educational Research Journal*, Vol. 40(1), pp. 281-310.

These findings support the assumption that embedding metacognitive guidance within co-operative settings is most effective for enhancing mathematical discourse, mathematics achievement, transfer of knowledge and mathematical creativity as reflected in students' explanations. When conditions prevent the implementation of co-operative settings, students exposed to metacognitive guidance in individualised settings still outperform their counterparts who were not exposed to metacognitive intervention. Similar findings were also reported by Cardelle-Elawar (1995) in a study that explored the role of metacognitive self-addressed questions on mathematics achievement of lower-achieving students in the third to eighth grades.

However, the implementation of metacognitive intervention in small groups was not always found to be beneficial. It depends on the task and the quality of the discourse in the small group (Artzt and Armour-Thomas, 1992). Complex tasks are more sensitive to metacognitive intervention than routine ones. Furthermore, Crook and Beier (2010) showed that learning for retention is superior when university students are trained to implement metacognitive processes individually rather than in dyads.

Which of the self-directed metacognitive questions are needed and for whom?

Almost all metacognitive interventions use self-directed questioning as a means to facilitate the application of metacognitive processes (see Chapter 4). These questions refer to understanding the problem, integrating new and existing knowledge, suggesting solution strategies (e.g. planning and monitoring), executing the solution, and evaluating it. An interesting question is whether they are all necessary in order to enhance metacognition and mathematics achievement. This question receives further relevance given the strong tendency in the educational literature, as well as in practice, to emphasise the importance of teaching strategies in themselves, particularly for lower achievers whose repertoire of learning strategies is deficient (e.g. NCTM, 2000). This tendency might make the other self-directed metacognitive questions superfluous.

To address this question, Mevarech (1999) compared the changes in mathematical reasoning of seventh graders who studied word problems using all four types of IMPROVE metacognitive questions (comprehension, connection, strategic and reflection), students who used only the strategic questions, and "traditional" instruction with no metacognitive guidance.

While in general, the IMPROVE group significantly outperformed the other two groups, surprisingly no significant differences were found among lower achievers between the group using only strategic questions and the control group, and in some cases, the group using strategic questions scored lower than the control group. This was not the case for the higher achievers: the IMPROVE students outperformed the "strategic group", who in turn outperformed the control group. For example, on the transfer tasks, 25% of the IMPROVE lower achievers succeeded in solving those tasks, compared with only 12% of the lower achievers in the control group, while none of the lower achievers in the strategic group were able to transfer their knowledge to the new problems. It is possible that there is little value in teaching the use of strategic questions to lower achievers, without training these students to also make connections and reflect on their solution process. If students do not understand when and why the strategies are of use, they may try to memorise the strategies, without being able to apply them in solving new problems. This explanation is indirectly supported by Schoenfeld (1989) who indicated that many students believe that succeeding in mathematics is based on their ability to recall.

When should metacognitive guidance be provided?

Is it better to implement the metacognitive instruction at the beginning of the solution process, only at the end, or during the solution process? So far, this interesting question has received little attention in the educational literature. Kapa (2001) assessed this issue in the area of algebra problem solving, and Michalsky, Mevarech and Haibi (2009) studied it in the context of science education.

Kapa (2001) designed computerised software for assisting eighth grade students to solve algebra word problems. The software provided different metacognitive hints according to the solution stage: during and at the conclusion of the solution, only during the solution, or only at the conclusion. Students were randomly assigned into one of four groups according to the type of metacognitive hints that were provided by the software, and to a control group that did not receive any metacognitive support. Kapa reported that students who received metacognitive guidance during the solution outperformed those who received metacognitive guidance only at the conclusion of the solution, whereas the control group who did not receive any computerised feedback attained the lowest mean score.

In the area of science education, Michalsky, Mevarech and Haibi (2009) provided metacognitive training during the reading of scientific texts. The examined outcomes were science literacy, domain-specific knowledge and metacognitive awareness. Fourth graders were assigned into one of four groups. One received metacognitive instruction at the beginning, one during, and one at the end of reading the scientific texts; the fourth group did not receive any explicit metacognitive instruction and served as a control group. Surprisingly, the findings indicate that the group who was exposed to IMPROVE immediately after completing the reading of the scientific texts outperformed the group who received the metacognitive guidance before the reading, who in turn outperformed the group who received the metacognitive guidance during the reading; the control group performed at the lowest level on all variables (see Figures 5.14 to 5.19). Similar findings were recently reported for lower secondary school students (Mevarech, Halperin and Vaserman, 2014). There is reason to suppose that the provision of metacognitive guidance during the reading overloaded the demands on the students who had to cope simultaneously with reading comprehension, understanding the scientific content, and applying the metacognitive hints. This hypothesis needs further examination.

Figure 5.14. **Effect of metacognitive guidance on overall science literacy**

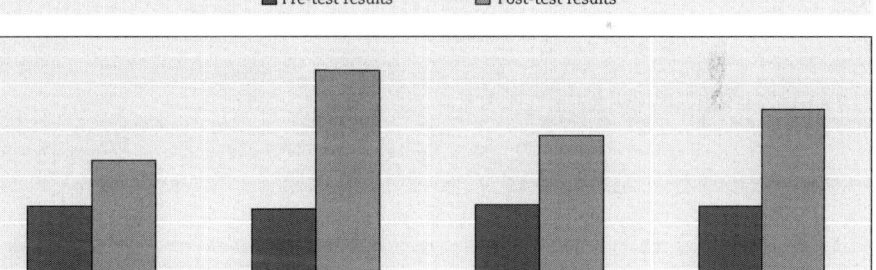

StatLink http://dx.doi.org/10.1787/888933149035

Source: Michalsky, Mevarech and Haibi (2009), "Elementary school children reading scientific texts: Effects of metacognitive instruction", *Journal of Educational Research*, Vol. 102(5), pp. 363-376.

Figure 5.15. **Effect of metacognitive instruction on describing phenomena**

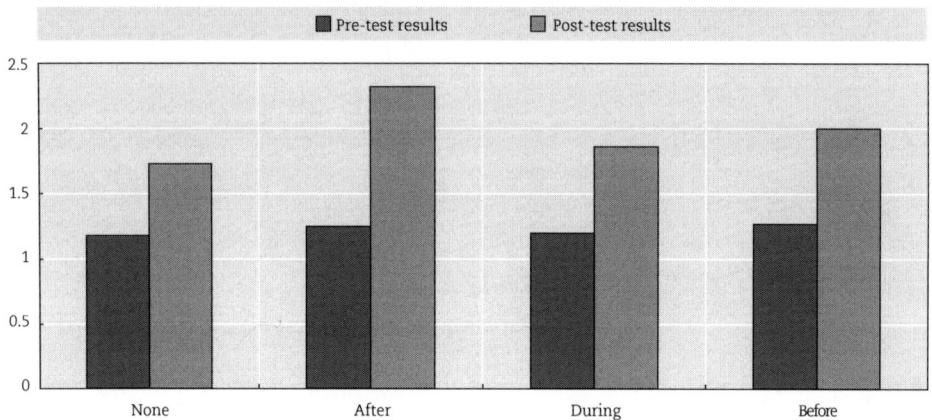

StatLink ⚙ http://dx.doi.org/10.1787/888933149040

Source: Michalsky, Mevarech and Haibi (2009), "Elementary school children reading scientific texts: Effects of metacognitive instruction", *Journal of Educational Research*, Vol. 102(5), pp. 363-376.

Figure 5.16. **Effect of metacognitive guidance on formulating hypotheses**

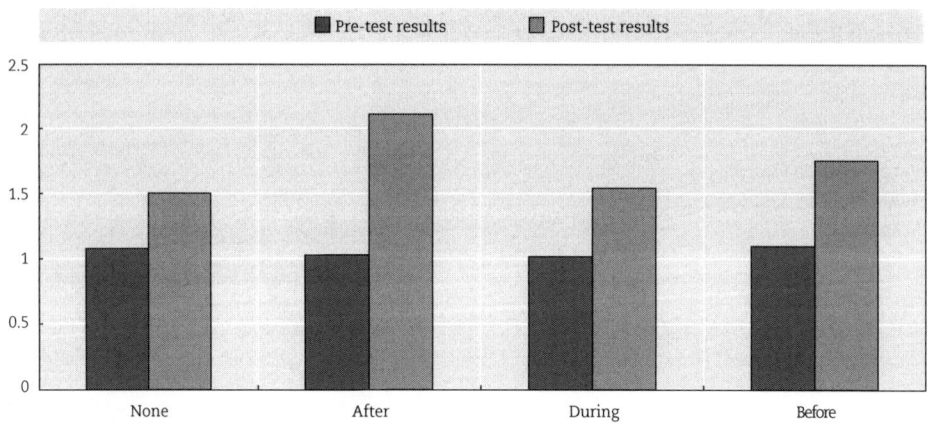

StatLink ⚙ http://dx.doi.org/10.1787/888933149056

Source: Michalsky, Mevarech and Haibi (2009), "Elementary school children reading scientific texts: Effects of metacognitive instruction", *Journal of Educational Research*, Vol. 102(5), pp. 363-376.

Figure 5.17. **Effect of metacognitive guidance on identifying results (dependent variables)**

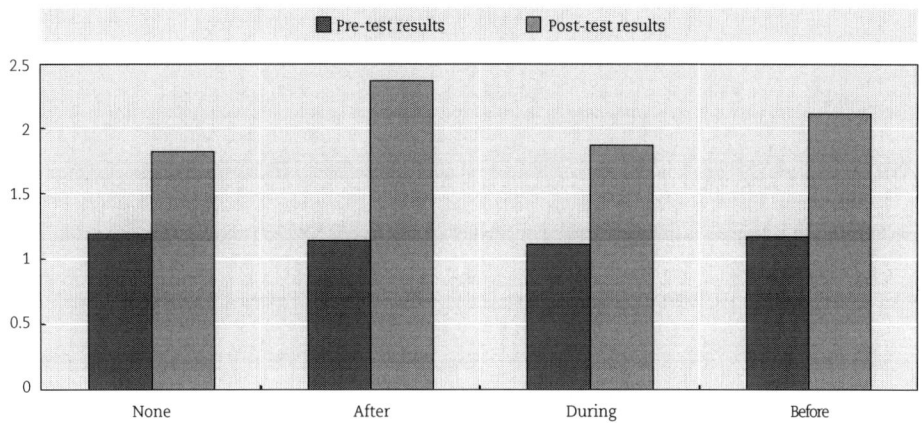

Source: Michalsky, Mevarech and Haibi (2009), "Elementary school children reading scientific texts: Effects of metacognitive instruction", *Journal of Educational Research*, Vol. 102(5), pp. 363-376.

Figure 5.18. **Effect of metacognitive guidance on identifying causes (independent variables)**

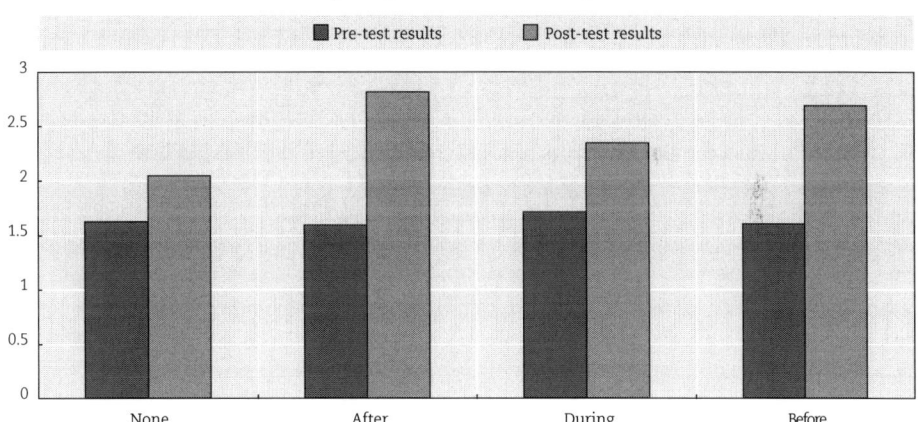

Source: Michalsky, Mevarech and Haibi (2009), "Elementary school children reading scientific texts: Effects of metacognitive instruction", *Journal of Educational Research*, Vol. 102(5), pp. 363-376.

Figure 5.19. **Effect of metacognitive guidance on reporting results and drawing conclusions**

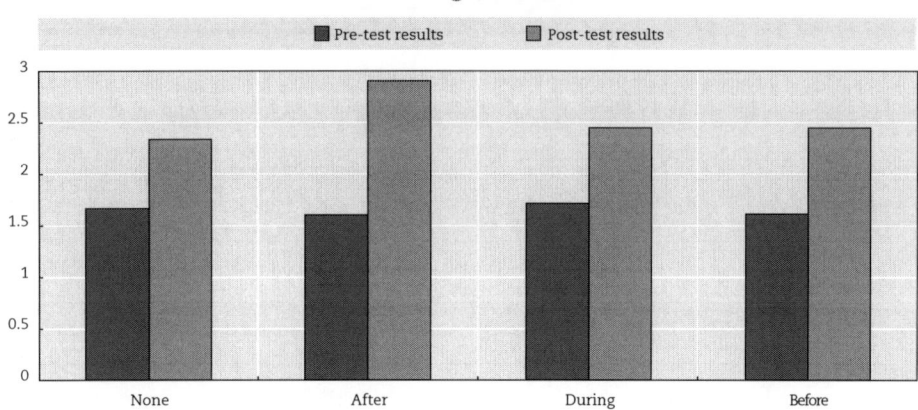

StatLink http://dx.doi.org/10.1787/888933149086

Source: Michalsky, Mevarech and Haibi (2009), "Elementary school children reading scientific texts: Effects of metacognitive instruction", *Journal of Educational Research*, Vol. 102(5), pp. 363-376.

Is metacognitive instruction in a single learning domain enough?

This issue is particularly relevant to the promotion of CUN problem solving, because these tasks require a number of different competencies, including those that are not directly related to the solution of routine mathematics problems (e.g. reading comprehension, making inferences or drawing conclusions). To address this issue, Kramarski, Mevarech, and Lieberman (2001) compared the effects of multilevel metacognitive training (MMT) which implemented IMPROVE in both maths and English as a foreign language classrooms, with uni-level metacognitive training (UMT), where the metacognitive training was implemented only in math classrooms, and a "NoMeta" control group whose students were not exposed to metacognitive instruction. For the purposes of this study, the principles of IMPROVE were adapted to the needs of studying English as a foreign language, and teachers in both the maths and English classrooms modelled the common use of metacognition in solving maths problems and in reading comprehension.

The findings suggest that multilevel metacognitive training enhances mathematics achievement and reasoning more than uni-level training, whereas the group with no metacognitive instruction attained the lowest scores on all variables. The gains of the MMT group were mostly evident on the solution of the Pizza task (see Box 1.1), an authentic transfer task that students had not had an opportunity to solve beforehand, and on metacognitive skills. Asking students to articulate their reasoning showed that the MMT group outperformed the UMT group on all four criteria for analysing students' discourse: referring to all data, organising information, processing information and drawing conclusions (Figure 5.20).

Figure 5.20. **Mathematics achievements on the Pizza task by learning conditions**

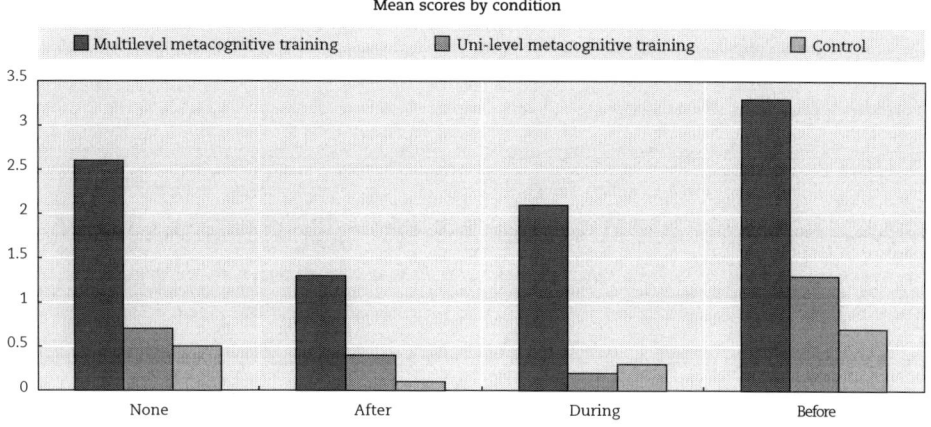

Mean scores by condition

■ Multilevel metacognitive training   ▨ Uni-level metacognitive training   ☐ Control

StatLink ⇒ http://dx.doi.org/10.1787/888933149090

Source: Kramarski, Mevarech & Lieberman (2001), "Effects of multilevel versus unilevel metacognitive training on mathematical reasoning", *The Journal of Educational Research*, Vol. 94(5), pp. 292-300.

At least three reasons may explain these findings. First, it is possible that a double metacognitive "dose" is more effective than a single dose. Second, according to constructivist theories, knowledge is retained and understood through elaborations and connections of different pieces of information (Wittrock, 1986). The likelihood of constructing connections between different types of information is greater in MMT than in UMT or NoMeta. Finally, because MMT teachers encouraged students to transfer knowledge and strategies from maths to English and vice versa, the MMT students were more willing than the UMT or the NoMeta student to work on the transfer task and implement the metacognitive processes even when they were not explicitly asked to do so. Investigating the impact of metacognitive instruction simultaneously in several domains and examining the differential effects on each domain is still an open issue that merits future research (Zimmerman and Schunk, 2011).

To sum up, researching the preferred conditions for implementing metacognitive pedagogies indicates that: 1) incorporating these methods within co-operative settings is more effective than in individualised settings, but in both cases, the inclusion of metacognitive instruction is more effective than no metacognitive instruction; 2) applying the full set of four self-addressed questioning suggested by IMPROVE is more effective than using only the comprehension and strategic questioning; and 3) the evidence for when best to administer the metacognitive scaffolding is inconclusive: while some studies reported higher scores when students were exposed to the metacognitive support during the problem-solving process, another reported greater effects when the metacognitive scaffolding was implemented at the end of the study.

Conclusion

The principles emerging from empirical studies on the effects of metacognitive pedagogies have important implications for their implementation. These are:

• Metacognitive pedagogies are important

The naïve approach that learners are passive absorbers of information has been replaced by constructivist theories assuming that learners have to be active builders of information. Since teaching methods powerfully influence how students build their knowledge, it is extremely important to implement those methods that have been found to be effective. IMPROVE and other metacognitive pedagogies have been proven to positively affect students' learning.

• Consider the quality of the different metacognitive pedagogies

Although the metacognitive pedagogies studied are ecologically valid in that the studies were implemented in real classroom situations, the specific pedagogies and the duration of the implementation vary. IMPROVE is an effective metacognitive instructional method. The generic metacognitive self-addressed questions encourage students to plan, monitor, control and reflect on the solution processes. IMPROVE is also one of the few methods that has been implemented over a full academic year and its lasting effects have been assessed through delayed tests and in high-stakes situations.

• Metacognitive pedagogies enhance students' performance of CUN tasks

Innovation-driven societies recognise the importance of focusing on CUN tasks. Given that IMPROVE trains students to plan, monitor, control and reflect on the solution processes by using comprehension, bridging, strategic, and reflective self-addressed questioning, it is not surprising that IMPROVE enhances students' solution of CUN and authentic tasks.

• Metacognitive pedagogies have positive effects on students' solutions of arithmetic, algebra and geometry problems

Many students need the metacognitive scaffolding in order to attain mastery of routine problem solving. The positive effects of IMPROVE and other similar metacognitive pedagogies have been found largely at all grade levels. The positive effects were evident in the areas of arithmetic and algebra as well as geometry.

• Consider the best conditions for implementing metacognitive pedagogies

The large number of studies examining the effects of metacognitive pedagogies enables us to identify the preferred conditions for implementation. Studies have shown that embedding IMPROVE in co-operative learning is more effective than in individualised learning, and that these two conditions are significantly more effective in enhancing mathematics learning than either co-operative or individualised settings with no metacognitive scaffolding. Research has also indicated that using all four types of self-addressed metacognitive questions is more effective than using only the strategic questions, particularly for lower-achieving students.

• Consider developmental issues in implementing metacognitive scaffolding

It is important to fit the teaching method to the students' age and their developmental level. At certain age, playing with concrete objects might be beneficial, whereas at other ages enhancing abstract thinking is more appropriate. IMPROVE has been implemented in all grade levels: kindergartens, primary schools, secondary schools and universities, modifying the metacognitive scaffolding to the student age, but using the same principles. Early exposure to metacognitive scaffolding might train learners to also use it later, in life-long learning situations. This issue requires further research.

• Incorporate metacognitive pedagogies in multiple disciplines simultaneously

Naturally, studies that aimed at enhancing mathematics achievement examined metacognitive pedagogies that were implemented only in mathematics classrooms. However, implementing metacognitive scaffolding via IMPROVE in both mathematics and English classes have additional benefits because students can generalise the use of the metacognitive processes beyond a specific domain.

• Build a strong data base for evidence-based policy making

There is no question about the importance of evidence-based policy making. It applies to medicine, economy, education, etc. Policy makers need to be informed about the benefits of metacognitive pedagogies and the hard data to support their effectiveness. They also have to be aware of the pitfalls that might be associated with these advanced pedagogies. Continuing to accumulate data will enrich our understanding of how learning occurs and will enable policy makers to draw valid conclusions. As the field of metacognition continues to progress, researching metacognition and its relationship to learning might lead to the design and wide implementation of metacognitive pedagogies for developing literate citizens, as well as for enhancing routine and CUN problem solving at different age groups and in various contexts.

References

Adibina, A. and I.J. Putt (1998), "Teaching problem solving to year 6 students: A new approach", *Mathematics Education Research Journal*, Vol. 10 (3), pp. 42-58.

Alin, R. (2012), "Teaching linear measurement in the Israeli kindergarten curriculum", in T. Papatheodorou and J. Moyles (eds.), *Cross-Cultural Perspectives on Early Childhood*, SAGE Publications Ltd, London, pp. 115-130.

Anderson, N.J. (2002). "The role of metacognition in second language teaching and learning", ERIC Digest, ERIC Clearinghouse on Languages and Linguistics, Washington, DC, *www.ericdigests.org/2003-1/role.htm*.

Artzt, A.F. and E. Armour-Thomas (1992), "Development of a cognitive-metacognitive framework for protocol analysis of mathematical problem solving in small groups", *Cognition and Instruction*, Vol. 9(2), pp. 137-175.

Blair, C. (2002), "School readiness: Integrating cognition and emotion in a neurobiological conceptualization of children's functioning at school entry", *American Psychologist*, Vol. 57(2), pp. 111–127.

Cardelle-Elawar, M. (1995), "Effects of metacognitive instruction on low achievers in mathematics problems", *Teaching and Teacher Education*, Vol. 11(1), pp. 81-95.

Cohors-Fresenborg, E. et al. (2010), "The role of metacognitive monitoring in explaining differences in mathematics achievement", *ZDM International Journal on Mathematics Education*, Vol. 42(2), pp. 231-244.

Crook, A., E. and M.E. Beier (2010), "When training with a partner is inferior to training alone: The importance of dyad type and interaction quality", *Journal of Experimental Psychology*, Vol. 16(4), pp. 335-348.

Davis, A. (2009) "So I'm done because I'm confused now: Measuring metacognition in Elementary Algebra community college students", PhD Thesis, UCLA.

De Corte, E., L. Verschaffel and P. Op 't Eynde (2000), "Self-regulation: A characteristic and a goal of mathematics education", in M. Bockaerts, P.R. Pintrich and M. Zeidner (eds.), *Handbook of Self-Regulation*, Academic Press, San Diego, CA, pp. 687-726.

Desoete, A. (2009), "Metacognitive prediction and evaluation skills and mathematical learning in third-grade students", *Educational Research and Evaluation*, Vol. 15, No. 5, pp. 435-446.

Dignath, C. and G. Buettner (2008), "Components of fostering self-regulated learning among students: A meta-analysis on intervention studies at primary and secondary school level", *Metacognition Learning*, Vol. 3, pp. 231-264.

Dignath, C., G. Buettner, and H.P. Langfeldt (2008), "How can primary school students learn self-regulated learning strategies most effectively? A meta-analysis on self-regulation training programmes", *Educational Research Review*, Vol. 3(2), pp. 101-129.

Edwards, T.G. (2008), "Reflective assessment and mathematics achievement by secondary at-risk students in an alternative secondary school setting", EdD thesis, Seattle Pacific University.

Elliott, A. (1993), "Metacognitive teaching strategies and young children's mathematical learning", working paper presented at the Australian Association for Research in Education Conference, Fremantle, WA, 22-25 November.

EU Council (2002), "Council resolution of 27 June 2002 on lifelong learning", *Official Journal of the European Communities*, C 163(1).

Frye, S.M. (1989), "The NCTM standards: Challenges for all classrooms", *Mathematics Teacher*, Vol. 2, pp. 313-317.

Garofalo, J. and F. Lester (1985), "Metacognition, cognitive monitoring and mathematical performance", *Journal for Research in Mathematics Education*, Vol. 16(3), pp. 63-176.

Hattie, J.A. (1992), "Measuring the effects of schooling", *Australian Journal of Education*, Vol. 36(1), pp. 5-13.

Hattie, J.A., J. Biggs and N. Purdie (1996), "Effects of learning skills interventions on student learning: A meta-analysis", *Review of Educational Research*, Vol. 66(2), pp. 99-136.

Hauptman, H. (2010), "Enhancement of spatial thinking with Virtual Spaces 1.0", *Computers and Education*, 54(1), pp. 123-135.

Hinton, C. and K. Fischer (2010), "Learning from the developmental and biological perspective", in H. Dumont, D. Istance and F. Benavides (eds.), *The Nature of Learning: Using Research to Inspire Practice*, Educational Research and Innovation, OECD Publishing, Paris, *http://dx.doi.org/10.1787/9789264086487-7-en*.

Hurme, T. Palonen and S. Jarvela (2006), "Metacognition in joint discussions: An analysis of the patterns of interaction and the metacognitive content of the networked discussion in mathematics", *Metacognition and Learning*, Vol. 1, pp. 181-200.

Kapa, E. (2001), "A metacognitive support during the process of problem solving in a computerized environment", *Educational Studies in Mathematics*, Vol. 47(3), pp. 317-336.

King, A. (1998), "Transactive peer tutoring: Distributing cognition and metacognition", *Educational Psychology Review*, Vol. 10(1), pp. 57-74.

Koriat, A. (2008), "Easy comes, easy goes? The link between learning and remembering and its exploitation in metacognition", *Memory and Cognition*, Vol. 36(2), pp. 416-428.

Kramarski, B. (2011), "Assessing self-regulation development through sharing feedback in online mathematical problem solving discussion", in G. Dettori and D. Persico (eds.), *Fostering SelfRregulated Learning through ICT*, IGI Global, pp. 317-247.

Kramarski, B. (2008), "Self-regulation in mathematical e-learning: Effects of metacognitive feedback on transfer tasks and self-efficacy", in A.R. Lipshitz and S.P. Parsons (eds.), *E-Learning: 21st Century Issues and Challenges*, Nova Science Publisher, New York, pp. 83-96.

Kramarski, B. and Z.R. Mevarech (2003), "Enhancing mathematical reasoning in the classroom: Effects of cooperative learning and metacognitive training", *American Educational Research Journal*, Vol. 40(1), pp. 281-310.

Kramarski, B., Z.R. Mevarech and M. Arami (2002), "The effects of metacognitive training on solving mathematical authentic tasks", *Educational Studies in Mathematics*, Vol. 49, pp. 225-250.

Kramarski, B., Z.R. Mevarech and A. Lieberman (2001), "Effects of multilevel versus unilevel metacognitive training on mathematical reasoning", *The Journal of Educational Research*, Vol. 94(5), pp. 292-300.

Kramarski. B., I. Weiss and I. Kololshi-Minsker (2010), "How can self-regulated learning support the problem solving of third-grade students with mathematics anxiety?", *ZDM International Journal on Mathematics Education*, Vol. 42(2), pp. 179-193.

Lajoie, S., et al. (2013), "Technology-rich tools to support self-regulated learning and performance in medicine" in R. Azevedo and V. Aleven (eds.), *International Handbook of Metaocnition and Learning Technologies*, Springer International Handbooks of Education, New York.

Leahey, E. and G. Guo (2001), "Gender differences in mathematics trajectories", *Social Forces*, Vol. 80(2), pp. 713-732.

Lester, F.K. (1994), "Musings about mathematical problem-solving research: 1970-1994", *Journal for Research in Mathematics Education*, Vol. 25(6), pp. 660-675.

Lovett. M.C. (2008), "Teaching metacognition", *http://net.educause.edu/upload/presentations/ELI081/FS03/Metacognition-ELI.pdf*

Lucangeli, D. and C. Cornoldi, C. (1997), "Mathematics and metacognition: What is the nature of the relationships?" *Mathematical Cognition*, Vol. 3(2), pp. 121-139.

McCabe, J. (2011), "Metacognitive awareness of learning strategies in undergraduates", *Memory and Cognition*, Vol. 39(3), pp. 462-476.

Mevarech, Z.R. (1999), "Effects of metacognitive training embedded in cooperative settings on mathematical problem solving", *Journal of Educational Research*, Vol. 92(4), pp. 195-205.

Mevarech, Z.R., and C. Amrany (2008), "Immediate and delayed effects of meta-cognitive instruction on regulation of cognition and mathematics achievement", *Metacognition Learning*, Vol. 3, pp. 147-157.

Mevarech, Z.R. and S. Fridkin (2006), "The effects of IMPROVE on mathematical knowledge, mathematical reasoning and meta-cognition", *Metacognition Learning*, Vol. 1, pp. 85-97.

Mevarech, Z.R., L. Gold, R. Gitelman and A. Gal-Fogel (2013), "Judgment of learning under different conditions: What works and what does not work?", Paper presented at 15th Biennial EARLI Conference for Research on Learning and Instruction, Munich, 27-31 August.

Mevarech, Z.R., C. Halperin and S. Vaserman (2014), "Reading scientific texts: the effects of metacognitive scaffolding on students' science literacy", 6th World Conference on Educational Sciences, Malta, 6-9 August.

Mevarech, Z.R. and M. Hillel (2012), "How and to what extent can children's metacognition be enhanced during mathematics problem solving?", in *Metacognition 2012 – Proceedings of the 5th Biennial Meeting of the EARLI Special Interest Group 16 Metacognition*, Milan, 5-8 September.

Mevarech, Z.R. and B. Kramarski (2003), "The effects of metacognitive training versus worked-out examples on students' mathematical reasoning", *British Journal of Educational Psychology*, Vol. 73(4), pp, 449-471.

Mevarech, Z.R. and B. Kramarski (1997), "IMPROVE: A multidimensional method for teaching mathematics in heterogeneous classrooms", *American Educational Research Journal*, Vol. 34(2), pp. 365-395.

Mevarech, Z.R., A. Tabuk and O. Sinai (2006), "Metacognitive instruction in mathematics classrooms: Effects on the solution of different kinds of problems", in A Desoete and M.V.J. Veenman (eds.), *Metacognition in Mathematics Education*, Nova Science Publishers, New York, pp. 70-78.

Mevarech, Z.R., S. Terkieltaub, T. Vinberger and V. Nevet (2010), "The effects of meta-cognitive instruction on third and sixth graders solving word problems", ZDM *International Journal on Mathematics Education*, Vol. 42(2), pp. 195-203.

Michalsky, T., Z.R. Mevarech and L. Haibi (2009), "Elementary school children reading scientific texts: Effects of metacognitive instruction", *Journal of Educational Research*, Vol. 102(5), pp. 363-376.

Mueller, J. (2012), "What is authentic assessment?", Authentic Assessment Toolbox website, *http://jfmueller.faculty.noctrl.edu/toolbox/whatisit.htm*.

NCTM (National Council of Teachers of Mathematics) (2000), *Principles and Standards for School Mathematics*, NCTM, Reston, VA.

Neeman, A., and Kramarski (submitted), "Metacognitive intervention intended to promote self-regulation and mathematics discourse in kindergarten students", paper submitted to the Metacognitive Special Interest Group Meeting, Istanbul, Turkey.

OECD (2012), PISA 2012 *Draft Frameworks — Mathematics, Problem Solving and Financial Literacy,* OECD, Paris, *www.oecd.org/pisa/pisaproducts/pisa2012draftframeworks-ma thematicsproblemsolvingandfinancialliteracy.htm.*

OECD (2007), PISA 2006: *Science Competencies for Tomorrow's World: Volume 1: Analysis,* PISA, OECD Publishing, Paris, *http://dx.doi.org/10.1787/9789264040014-en.*

OECD (2003), *The PISA 2003 Assessment Framework: Mathematics, Reading, Science and Problem Solving Knowledge and Skills,* Education and Skills, OECD Publishing, Paris, *http://dx.doi.org/10.1787/9789264101739-en.*

OECD/UNESCO Institute for Statistics (2003), *Literacy Skills for the World of Tomorrow: Further Results from PISA 2000,* PISA, OECD Publishing, Paris, *http://dx.doi. org/10.1787/9789264102873-en.*

Panaoura, A., A. Demetriou and A. Gagatsis. (2009), "Mathematical Modeling, self-representations and self-regulation", in *Proceedings of CERME 6,* Lyon, France, 28 January – 2 February, *www.inrp.fr/editions/cerme6.*

Pennequin, V., O. Sorel, I. Nanty and R. Fontaine (2010), "Metacognition and low achievement in mathematics: The effect of training in the use of metacognitive skills to solve mathematical word problems", *Thinking & Reasoning,* Vol. 16(3), pp. 198-220.

Perry, N.E., L. Phillips and J. Dowler (2004), "Examining features of tasks and their potential to promote self-regulated learning", *Teachers College Record,* Vol. 106(9), pp. 1854-1878.

Perry, N.E., K.O. VandeKamp, L.K. Mercer and C.J. Nordby (2002), "Investigating teacher-student interactions that foster self-regulated learning", *Educational Psychologist,* Vol. 37(1), pp. 5-15.

Pintrich, P.R. (2000), "Multiple goals, multiple pathways: The role of goal orientation in learning and achievement", *Journal of Educational Psychology,* Vol. 92(3), pp. 544-555.

Ragosta, P. (2010), "The effectiveness of intervention programs to help college students to acquire self-regulated learning strategies: A meta-analysis", Ph.D Thesis, City University, New York.

Schoenfeld, A.H. (1992), "Learning to think mathematically: Problem solving, metacognition, and sense-making in mathematics", in D.A. Grouws, (ed.), *Handbook of Research on Mathematics Teaching,* MacMillan Publishing, New York, pp. 334-370.

Schoenfeld, A.H. (1989), "Problem solving in context(s)", in R. Charles and E. Silver (eds.), *The Teaching and Assessing of Mathematical Problem Solving,* NCTM , Reston, VA, pp. 82-92.

Schraw, G. (1998), "Promoting general metacognitive awareness", *Instructional Science,* Vol. 26, pp. 113-125.

Schraw, G., K.J. Crippen and K. Hartley (2006), "Promoting self-regulation in science education: Metacognition as part of a broader perspective on learning", *Research in Science Education*, Vol. 36, pp. 111-139.

Schraw, G. and R.S. Dennison (1994), "Assessing metacognitive awareness", *Contemporary Educational Psychology*, Vol. 19, pp. 460-475.

Schraw, G. and D. Moshman (1995), "Metacognitive theories", *Educational Psychology Review*, Vol. 7(4), pp. 351-371.

Schwonke, R., A. Ertelt, C. Otieno, A. Renkl, V. Aleven and R. Salden (2013), "Metacongitive support promotes an effective use of instructional resources in intelligent tutoring", *Learning and Instruction*, Vol. 23, pp. 136-150.

Shayer, M. and P.S. Adey, (1993), "Accelerating the development of formal thinking in middle and high school students IV: Three years after a two-year intervention", *Journal of Research in Science Teaching*, Vol. 30(4), pp. 351-366.

Stasi, G.M. (2005), "Differential effects of a content-oriented metacognitive instructional program and a process-oriented metacognitive instructional program", PhD Thesis, Illinois Institute of Technology.

Stillman, G. and Z.R. Mevarech (2010), "Metacognitive research in mathematics education: from hot topic to mature field", *ZDM International Journal on Mathematics Education*, Vol. 49(2), pp. 145-148.

Subocz, S.L. (2007), "Attitudes and performance of community college students receiving metacognitive strategy instruction in mathematics courses", PhD Thesis, Capella University.

TIMSS (Third International Mathematics and Science Study) (1997), *Performance Assessment in IEA's Third International Mathematics and Science Study (TIMSS)*, TIMSS International Study Center, Boston College, MA.

Veenman, M.V.J., B.H.A.M. Van Hout-Wolters and P. Afflerbach (2006), "Metacognition and learning: Conceptual and methodological considerations", *Metacognition and Learning*, Vol. 1, pp. 3-14.

Verschaffel, L. (1999), "Realistic mathematical modeling and problem solving in the upper elementary school: Analysis and improvement", in J.H.M Hamers, J.E.H Van Luit and B. Csapo (eds.), *Teaching and Learning Thinking Skills. Context of Learning*, Swets and Zeitlinger, Lisse, pp. 215-240.

Verschaffel, L., B. Greer and E. De Corte (2000), *Making Sense of Word Problems*, Swets and Zeitlinger, Lisse.

Weiss, I. and J. Pasley (2004), "What is high-quality instruction?" *Educational Leadership*, Vol. 61(5), pp. 24-28.

Whitebread, D. and P. Coltman (2010), "Aspects of pedagogy supporting metacognition and self-regulation in mathematical learning in young children: Evidence from an observational study", *ZDM International Journal on Mathematics Education*, Vol. 42(2), pp. 163-178

Whitebread, D. et al. (2009), "The development of two observational tools for assessing metacognition and self-regulated learning in young children", *Metacognition and Learning*, Vol. 4(1), pp. 63-85.

Winne, P.H. (1995), "Inherent details in self-regulated learning", *Educational Psychologist*, Vol. 30(4), pp. 173-187.

Wittrock, M.C. (1986), "Students' thought processes", in M.C. Wittrock (ed.), *Handbook of Research on Teaching*, MacMillan, New York, pp. 297-314.

Yang, Kai-Lin (2012), "Structures of cognitive and metacognitive reading strategies use for reading comprehension of geometry proof", *Educational Studies in Mathematics*, Vol. 80, pp. 307-326

Yimer, A. and N.F. Ellerton (2010), "A five-phase model for mathematical problem solving: Identifying synergies in pre-service-teachers' metacognitive and cognitive actions", *ZDM International Journal on Mathematics Education*, Vol. 42, pp. 245-261.

Zimmerman, B.J. (2000), "Attainment of self-regulated learning: A social cognitive perspective", in M. Boekaerts, P. Pintrich and M. Zeidner (eds.), *Handbook of Self-Regulation*, Academic Press, Orlando, FL. pp. 13-39.

Zimmerman, B.J. and D. Schunk, (eds.) (2011), *Handbook of Self-Regulation of Learning and Performance*, Routledge, New York.

Chapter 6

The effects of metacognitive pedagogies on social and emotional skills

Emotion and cognition are inextricably linked in the brain. Social skills are essential to the process of learning and the evidence shows that metacognitive interventions designed to improve cognitive achievement can have a beneficial impact on affective factors such as motivation or anxiety. In addition, metacognitive methodologies can be adapted to promote social-emotional competencies among kindergarten pupils, primary and secondary school students, and adults. Combining the two approaches has an even greater impact on both social-emotional and cognitive achievements than either one on its own. Interventions that focus only on motivation or only on cognitive-metacognitive competencies are more effective than traditional instruction, but less effective than focusing on both motivation and metacognition.

Current research in neurosciences indicates that "emotion and cognition are inextricably linked in the brain. Particular components of the (learning) experience can usefully be labelled cognitive or emotional, but the distinction between the two is theoretical since they are integrated and inseparable in the brain" (Hinton and Fischer, 2010, p. 119).

These findings, surprising as they are, also pinpoint the role of social interaction in learning, showing "the essential social nature of human learning. The human brain is primed for social interaction. The brain is tuned to experience empathy, which intimately connects us to others' experiences... People use their brains to learn through social interactions and cultural context" (Hinton and Fischer, 2010, pp. 126 and 129, respectively). There are at least two reasons why parcelling the brain into cognitive and affective or emotional-social regions is inherently problematic: first, brain regions viewed as "affective" are also involved in cognition, and vice versa; and second, even more critically, cognition and emotion are integrated in the brain and actually the two systems (social-emotion and cognition) interact in important ways (e.g. Pessoa, 2008).

What is the neurological mechanism that converts external experience into emotions and/or social processes? Hinton and Fischer (2010) explain that the "learning experiences are translated into electrical and chemical signals that gradually modify connections between neurons" (p. 118). Since the neurons in the brain are organised in modules, a stimulus elicits a network of responses from different modules. The initial connection is temporary, and repeated activities eventually lead to long term changes that underlie the long term memory. In addition, neuroscientists have discovered "mirror neurons" that fire to stimulate others' experience (Dobbs, 2006). Hinton and Fischer (2010) further clarify: "when a child sees his or her mother build a tower of blocks, some of the same neurons in the child's brain fire as when the child builds a tower of blocks himself or herself. These mirror neurons are thought to be the neurological basis for empathy, and serve both bonding and learning" (p. 126).

Interestingly, even before recognising the changes in the brain caused by various experiences, psychologists and educators identified the characteristics of what they have called "meta-emotions" or "meta-experiences" that, as with metacognition, monitor, control and regulate human behaviour. Gottman, Katz and Hooven (1997) explained the term meta-emotion as emotion about emotion, analogous to metacognition (p.6). They define meta-emotion as "an organized and structured set of emotions and cognitions about emotions, both one's own emotions and the emotions of others" (p. 7). Efkelides (2006, 2011) went one step further looking at meta-experience (ME) during learning. According to her studies, meta-experiences refer to what a person is aware of and what she or he feels when coming across a task and processing the information related to it. Efkelides identified three ME categories: "feelings", "judgments or estimations", and "online task-specific knowledge". The first category relates to feelings of knowledge (FOK) as well as to feelings of success / failure, familiarity / difficulty, self-confidence, and satisfaction, both personal and related to specific tasks. Judgments refer to judgment of learning (JOL), source of memory information, estimate of efforts and estimate of time. Finally, the online task-specific knowledge denotes task features, and the procedures employed.

Efkelides (2006) theorises that metacognition and meta-experience act together in regulating one's learning. On the basis of her studies, we present in Figure 6.1 the relationships between metacognition (metacognitive knowledge and metacognitive skills) and metacognitive experiences.

Figure 6.1. **Relationship between metacognition and metacognitive experiences**

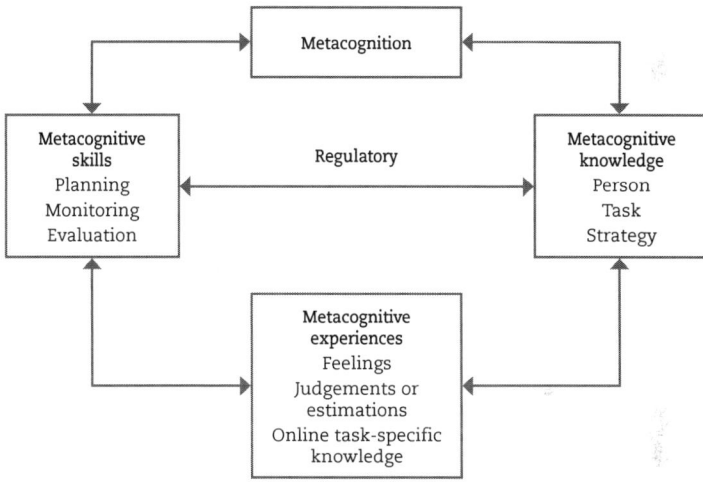

While Efkelides (2006, 2011) emphasises the roles of emotions in learning, others (e.g. Zins, Bloodworth, Weissberg and Walberg, 2007) added the essential contribution of social competencies to the process of learning. Zins et al. (2007) indicate that "social and emotional education involves teaching children to be self-aware, socially cognizant, able to make responsible decisions, and competent in self-management and relationship-management skills so as to foster their academic success" (p. 6). Thus, children need to be aware of themselves and others, give consideration to the situation and relevant norms, manage their emotions and behaviour, and possess social skills that enable them to learn effectively and collaborate with others. These skills and attitudes positively or negatively affect students' engagement in learning which in turn affects schooling outcomes. Box 6.1 specifies the social and emotional skills that are related to learning.

Box 6.1. Social-emotional components and skills

Self awareness:
- identifying and recognising emotions
- developing accurate self-perception and self-efficacy
- recognising strengths, needs and values

Social awareness:
- perspective taking: taking each other's points of views into account
- empathy
- appreciating diversity
- respect for others

Responsible decision making:
- problem identification and situation analysis
- problem solving
- evaluation and reflection
- personal, moral and ethical responsibility

Self management:
- impulse control and stress management
- self-motivation and discipline
- goal setting and organisational skills

Relationship management:
- communication, social engagement and building relationships
- working co-operatively
- negotiation, refusal and conflict management
- Help seeking and providing

The following 15 social skills are involved in promoting self-regulated learning:

1. recognising emotions in self and others
2. regulating and managing strong emotions (positive and negative)
3. recognising strengths and areas of need
4. listening and communicating accurately and clearly
5. understanding others' perspectives and sensing their emotions
6. respecting others and oneself and appreciating differences
7. identifying problems correctly
8. setting positive and realistic goals
9. problem solving, decision making and planning
10. approaching others and building positive relationships
11. resisting negative peer pressure
12. co-operating, negotiating and managing conflict non violently
13. working effectively in groups
14. seeking and giving help
15. showing ethical and social responsibility

Source: Based on ZINS et al. (2007).

Yet, children are not very skilled in regulating their social-emotional processes. Social-emotional skills do not develop spontaneously, particularly not in children at high risk of developing learning disabilities. Research has indicated that more explicit and intentional teaching is needed (Bredekamp and Copple, 1997). The preschool and early school-years would seem to be a strategic time for interventions aiming to facilitate social competencies and reduce aggressive behaviour before it develops into more permanent patterns. Nevertheless, it is "never too late" to intervene even at a later stage.

Can social-emotional skills be taught?

The answer is yes. For example, one of the teaching methods that has been designed for this purpose is the "social information-processing model" that is widely used to understand and foster social competencies (Crick and Dodge, 1994). The social information-processing model focuses directly on the cognitive / metacognitive processes. In general, this process consists of selecting a social goal, monitoring the environment, generating and selecting a strategy, implementing the strategy, evaluating its outcome, and deciding on subsequent action.

The social information-processing model outlines a six-step nonlinear process with various feedback loops linking children's social cognition and behaviour (Table 6.1).

Table 6.1. **Six-step information processing model**

Steps	Cognitive processes
Observation and encoding of relevant stimuli	Attending to and encoding non-verbal and verbal social cues, both external and internal.
Interpretation and mental representation of cues	Understanding what has happened during the social encounter, as well as the cause and intent underlying the interaction.
Clarification of goals	Determining what one's objective is for the interaction and how to put forth an understanding of those goals
Representation of situation is developed by accessing long	Comparing the interaction to previous situations stored in long-term memory and the previous outcomes of those interactions
Response decision/selection	
Behavioural enactment and evaluation	

Source: Based on Crick and Dodge (1994).

The model thus emphasises the importance of being aware of the social-emotional interactions that take place in the classroom, how to collect information about it, what to do with the information and reflecting on the obtained outcome. Since the steps described above are not linear, students are trained to move smoothly between the steps as the situation requires.

The social-information processing model is not the only model that proposes an intervention programme for fostering social-emotional competencies. The new findings showing that cognition and social-emotional processes work in tandem open new venues for educational researchers. Co-operative learning is the most distinctive instructional method to be based on these findings, providing evidence on how social interactions facilitate learning (e.g. Slavin, 2010). Yet, many questions remain open: is the enhancement of social-emotional competencies a prerequisite for attaining cognitive-metacognitive goals, or vice versa? Is student engagement in CUN tasks sufficient to promote social-emotional processes? What types of pedagogies (e.g. co-operative learning or metacognitive scaffolding) are appropriate for fostering social-emotional skills? Is learning in small groups in itself sufficient for facilitating social skills? Finally, are there robust findings regarding these practices? These are important questions because quite often emotions regulate learning, as in the case of mathematics anxiety which decreases the ability to learn and causes negative outcomes for many students. This chapter addresses some of these issues, assuming that social-emotional competencies are important outcomes by themselves for innovative societies.

Metacognitive pedagogies and their effects on social-emotional competencies

Studies on the effects of metacognitive pedagogies on social-emotional competencies can be classified into three major types: 1) studies focusing on cognition-metacognition but not explicitly on emotional skills assuming that by enhancing cognitive-metacognitive outcomes also emotional factors will improve; 2) studies focusing on emotions, attempting through that to increase cognitive achievements as well; and 3) studies focusing on both cognition-metacognition and emotion, on the grounds that both are needed in order to foster cognitive and emotional outcomes.

In all three types of studies, the dependent variables include various schooling outcomes: cognitive (e.g. routine and CUN problem solving), metacognitive and social-emotional (e.g. students' engagement, social skills and communication). Therefore, we will typically report on all these outcomes, addressing the important issue of whether or not a specific method not only improves social and emotional skills, but also leads to more effective learning. While there could sometimes be a trade-off, a powerful pedagogy would be one that fosters both outcomes simultaneously.

Furthermore, unless otherwise indicated, the studies reviewed below are based on quasi-experimental designs in which the reported differences between the experimental and control groups are statistically significant. Although in all the studies the outcomes considered include both emotional and cognitive components, this chapter concentrates on the studies which focused mainly on the emotional components, whereas those reported in the previous ones mainly focused on the cognitive-metacognitive outcomes. For example, studies on mathematical communication often reflect maths reasoning and therefore these studies are reviewed in Chapter 5.

Type I studies: the effects of achievement-focused interventions

Alleviating maths anxiety

Anxiety consists of cognitive, affective and behavioural components (Ziedner, 1998). The cognitive component refers to intrusive thoughts that pop into one's mind during learning but have no functional value in solving the cognitive task at hand. The affective component includes feelings of nervousness, tension and unpleasant physiological reactions to threatening situations. The behavioural component refers to a variety of avoidance or escape behaviours at various stages of the solution process. All these processes may be evident when students learn, whether individually or in groups (Hembree, 1990).

Mathematics often arouses negative emotions, such as anxiety or negative self-esteem. This phenomenon is so spread that it has been widely researched. Mathematics anxiety is defined as "feelings of tension and anxiety that interfere with the manipulation of numbers and the solving of mathematical problems in a wide variety of ordinary life and academic situations" (Richardson and Suinn, 1972, p. 551). Furthermore, empirical evidence (e.g. Pintrich, 2000; Schraw et al., 2006) indicates that negative feelings and low task expectations might prevent students from being engaged in the learning activity, and vice versa. Bad learning experiences often arouse negative feelings that in turn decrease mathematics achievement (Efklides, 2011). A meta-analysis study showed that once negative attitudes are formed they can be intractable, persisting into adulthood with far-reaching consequences such as maths avoidance (Hembree, 1990). Maths anxiety is evident so often that a large number of treatments have been suggested to cope with the phenomenon. Most of these treatments are effective in reducing maths anxiety and improving mathematics performance (Hembree, 1990).

To what extent, then, can metacognitive pedagogies reduce these bad feelings towards mathematics? If indeed emotions, cognition, and metacognition are complementary so that one triggers the other, there is reason to suppose that metacognitive pedagogies that enhance cognitive outcomes would result also in fostering social-emotional processes, such as those that relate to the learning of mathematics.

Kramarski et al. (2010) addressed this issue by investigating the effects of IMPROVE on mathematics anxiety and mathematics problem solving of 140 third grade Israeli students (lower and higher achievers). About half of the students were exposed to IMPROVE, and the others acted as a control group and received no direct metacognitive support. All students were administered a mathematics test and a questionnaire that focused on their self-perceived mathematics anxiety. Findings indicated that compared with the control group, the IMPROVE students showed greater gains in their mathematical problem-solving performance on basic, complex and transfer tasks (the mean differences between experimental and control groups were 11, 10 and 24 points, respectively). In addition, the IMPROVE students reported using metacognitive strategies more often, and self-reported a greater reduction in mathematics anxiety, as indicated by a lower level of negative thoughts and avoidance of bad feelings/ behaviour towards mathematics. All these differences were statistically significant. Figure 6.2 presents the mean scores on mathematics anxiety for higher and lower achievers by intervention as reported by Kramarski et al. (2010).

Figure 6.2. **Changes in maths anxiety for high and low achievers**

Mean scores by condition

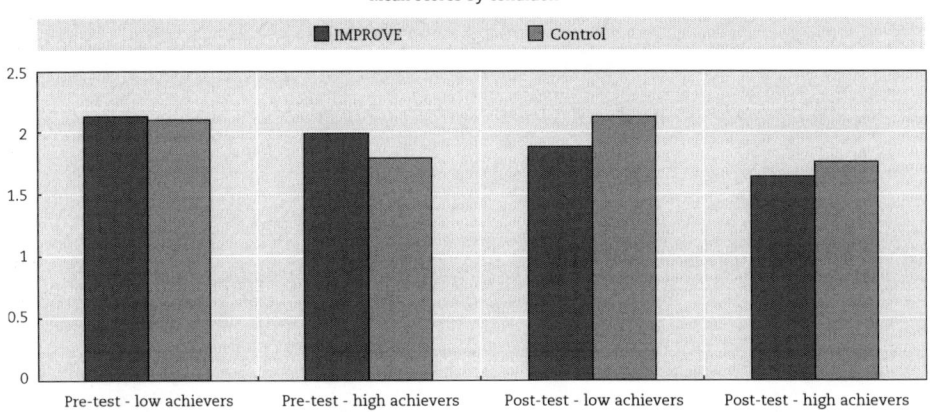

StatLink http://dx.doi.org/10.1787/888933149109

Note: Lower scores indicate less anxiety.

Source: Kramarski, Weiss & Koloshi-Minsker (2010), "How can self-regulated learning support the problem solving of third-grade students with mathematics anxiety?", ZDM International Journal on Mathematics Education, Vol. 42(2), pp. 179-193.

According to Figure 6.2, under the IMPROVE-condition, both lower and higher achievers reduced their level of maths anxiety. In the control group, only the higher achievers suffered less maths anxiety, while the lower achievers reported even greater anxiety than prior to the beginning of the study, probably because of the increased difficulty level of the learning unit. However, since all measurements are based on self-reports with no observation or use of think aloud methods, the findings might be biased. This issue merits future research.

Kramarski et al. concluded that implementing IMPROVE as originally designed (Mevarech and Kramarski, 1997) enabled third graders to simultaneously alleviate their maths anxiety, enhance their metacognitive strategies, and ability to solve basic, complex, and transfer tasks. This study design, however, does not answer the question of the extent to which the enhancement of socio-emotional competencies is a prerequisite for attaining cognitive and metacognitive goals or vice versa. It only shows that cognitive-metacognitive intervention, even if not explicitly focused on emotional skills, could alleviate mathematics anxiety in addition to increasing achievement on routine and CUN tasks.

While Kramarski et al. focused on primary school students, Shen (2009) investigated the effects of metacognitive pedagogies on mathematics anxiety, motivation, and mathematics achievement of university-level psychology students. Shen conducted a 2x2 study design in which one factor refers to emotional support (provided or not provided) and the other to cognitive-emotional support (provided or not provided). Emotional support was provided via exposing students to computer messages such as "… I was also an anxious student. I know you are feeling anxious now. I know what that's like when I had the same class last year." The cognitive-emotional factor included computer messages such as "this instructional module will help you to answer similar problems on the maths exam". The findings indicated that students who were exposed to the emotional support messages outperformed the other groups and had also lower levels of maths anxiety than those students who studied with no emotional support. No significant differences were found for the main effect of the cognitive-emotional provision. This study shows how a simple intervention can alleviate mathematics anxiety among college students.

Motivation and self-efficacy

The focus on mathematics anxiety, important as it is, is just a single dimension of the enhancement of socio-emotional skills. Based on positive psychology, many researchers claim that instead of alleviating negative emotions, educators have to explicitly foster positive emotions, particularly those related to students' engagement in learning, including intrinsic motivation and self-efficacy. Positive measures of socio-emotional skills therefore also need to be considered.

In this spirit, Mevarech, Michalsky, and Sasson (submitted) analysed the effects of IMPROVE on students' science literacy, and also on their self-efficacy and intrinsic motivation for studying science. The objective of this study was threefold: 1) to examine the effects of metacognitive pedagogy (IMPROVE) on students' science literacy in biology; 2) to explore the transfer of knowledge of this pedagogy to another science literacy domain (physics); and 3) to research the effects of this pedagogy on students' intrinsic motivation and self-efficacy.

The participants were ninth grade students (aged around 15 years old) in six science classes who studied science four times a week, of which one period was devoted to reading texts about biological research. All students were pre- and post-tested on biology literacy and then also on physics literacy (to measure transfer of knowledge). All the tasks in the examinations were selected from science literacy exams of the OECD Programme for International Student Assessment (PISA) or were designed by the authors according to the PISA framework. In addition, all students were pre- and post-tested on their intrinsic motivation and self-efficacy regarding the learning of science.

The six classes were randomly assigned into one of two conditions: a control group, who learned with no metacognitive intervention, and IMPROVE students who were guided to use self-addressed metacognitive questioning. Figure 6.3 presents the mean scores by time and treatments on science literacy (biology and physics) and affective outcomes (motivation and self-efficacy) as reported by Mevarech et al. (submitted).

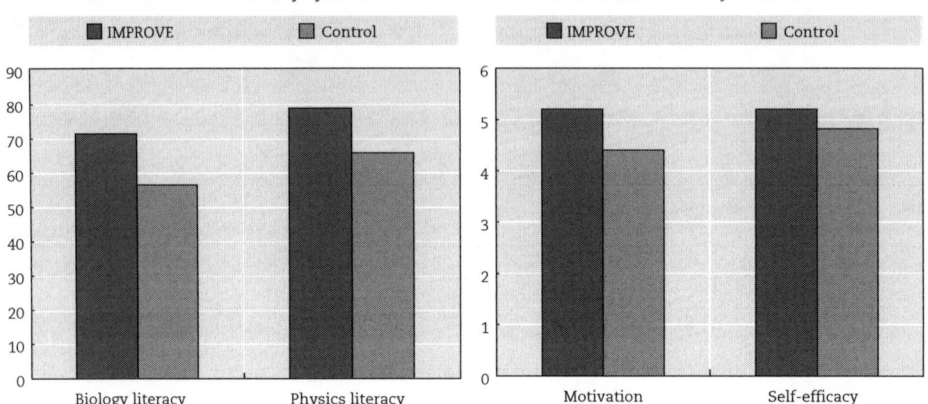

Figure 6.3. **Effects on scientific literacy, motivation and self-efficacy**

StatLink ⬚⬚⬚ http://dx.doi.org/10.1787/888933149113

Source: Mevarech, Michalski and Sasson (submitted), "Meta-cognition and science literacy: Immediate and transferred effects on science literacy, motivation, and self-esteem", paper submitted to the European Associate Research on Learning and Instruction (EARLI) Meeting, 2014.

The results indicated that while no significant differences were found between the groups prior to the beginning of the study, the IMPROVE students significantly outperformed the control group on science literacy in biology (Mean = 56.5 and 71.5; Standard Deviation = 25.5 and 24.0 for the control and experimental groups, respectively) and on the transfer tasks in physics (Mean = 65.9 and 79.0 Standard Deviation = 23.5 and 16.5 for the control and experimental groups, respectively). Similar results were also found for student engagement (intrinsic motivation and self-efficacy): the IMPROVE students were more motivated to study science (Mean = 4.4 and 5.20; Standard Deviation = 1.09 and .94 for the control and experimental groups, respectively), and the level of self-efficacy among the IMPROVE group was higher than that of the control group (Mean = 4.82 and 5.21; Standard Deviation = .94 and .79, respectively).

These findings indicate that IMPROVE has the potential to enhance motivation and self-esteem alongside facilitating science literacy in accordance with the PISA conceptualisation. In addition, the findings emphasise the role of metacognitive scaffolding in fostering students' ability to solve CUN physics tasks (transfer of knowledge).

Type II studies: using metacognitive pedagogies to promote social-emotional competencies

The importance of social-emotional skills, and the very fact that the cognitive and social-emotional systems interact with each other, have led many researchers to design metacognitive interventions to foster social-emotional skills. Although the programmes differ in their details, most of these interventions are based on the same principles as the metacognitive interventions described above. Students are trained to identify or recognise the problem, plan the strategy to be used, control and monitor the behaviour or solution, and reflect on the process and the outcomes. Furthermore, many of these programmes are implemented in small groups, and students are encouraged to articulate their thoughts and feelings while using self-addressed questions similar to those suggested by IMPROVE.

In this section we review several metacognitive programmes specially designed to foster students' social-emotional processes. These covered students in kindergarten, primary and secondary schools, and in colleges. With the exception of the kindergarten programme, all of them focused on achievement in mathematics, science or other domains as well as the social-emotional competencies.

DARE to be you: Tools for promoting educational success and social skills in kindergarten children

Webster-Stratton and Reid (2004, 2007) conducted a series of studies in which kindergarten children learned a seven-step process of problem solving. The steps were:

1. How am I feeling, and what is my problem? (Define the problem and feelings).

2. What is your solution?

3. What are some more solutions? (Brainstorm solutions or alternative choices).

4. What are the consequences?

5. What is the best solution? (Is the solution safe, fair, and does it lead to good feelings?).

6. Can I use my plan?

7. And finally how did I do? (Evaluate outcome and reinforce efforts).

These seven steps reflect the basic principles of metacognitive pedagogies: defining the problem, bridging, suggesting strategies and reflecting. This is not surprising, as all these programs focus on solving problems, even though the problems are different. Webster-Stratton and Reid (2004) used the Wally Problem Solving test which consists of thirteen situations to which the child has to respond (rejection, making mistakes, unjust treatment, victimisation, prohibition, loneliness, being cheated, disappointment, dilemma, adult disapproval, and attack). They assessed the effectiveness of the programme on the children's social-emotional skills through classroom observations of children and teachers in structured and unstructured settings. Based on these measures, they reported that the experimental group had higher post-intervention scores on positive coping skills, significant improvements in pro-social skills, and an overall reduction of problem behaviour. The control group, on the other hand, had no improvement in pro-social behaviour and no increase in problem behaviour. Several replication studies found similar results (Webster-Stratton and Reid, 2004).

The RULER programme for primary school students

In a similar vein, Brackett, Rivers, Reyes and Salovey (2012) conducted a study in which 273 fifth and sixth graders learned the RULER approach:

- *Recognise* emotions in oneself and in other people.

- *Understand* the causes and consequences of a wide range of emotions.

- *Label* emotions using a sophisticated vocabulary.

- *Express* emotions in socially appropriate ways.

- *Regulate* emotions effectively.

The effects were examined by using teachers' assessments based on the Behavioral Assessment System for Children (BASC) questionnaire (Reynolds and Kamphous, 1992, 2004). The teachers indicated the extent to which each child in their classrooms engaged in 148 different kinds of behaviour (e.g. give up easily, skips classes at school, studies with others, hyperactive, aggressive, creative, shows leadership). The findings revealed that compared with students in the comparison group, RULER students had higher year-end grades in English (the context in which RULER was implemented), but not in mathematics. In addition, RULER teachers scored their students' social and emotional competencies more highly (e.g. leadership, positive interactions with other students, persistence on tasks, creativity, no discipline problems).

Ornaghi et al. (2012) similarly investigated whether training primary school children (second graders, about seven years old) to regulate their emotions played a significant role in improving their emotional understanding and social-cognition abilities. During a two month intervention in small groups, the experimental group was involved in metacognitive conversations about the *nature* of emotions (e.g. identifying, reflecting and discussing emotions as expressed in words, faces, body language, etc.), the *external and internal* causes of the identified emotions (e.g. bridging and comprehension), and *regulation strategies* to deal with the emotions. The study focused on five emotions: fear, anger, sadness, guilt and happiness.

Results indicated that the training group significantly outperformed the control group on using language to describe emotions, comprehending emotions and situational comprehension ("How do I feel in different situations"). In addition, the intervention group developed a higher level of empathy than the control group: children in the experimental group were more inclined to put themselves in others' shoes, to recognise and understand the feelings of other people, and be emotionally involved in their feelings. Furthermore, the experimental group scored higher on the achievement test in the area in which the intervention was implemented (English), but not in mathematics. The authors concluded that teachers can use "metacognitive conversation" to encourage children to think and talk about their own and others' emotions and by that improve students' social understanding and socio-emotional competencies.

Affective Regulation Model: A modified version of IMPROVE

Another version of RULER was suggested by Tzohar-Rozen and Kramarski (in press). Based on the model proposed by Pintrich (2000), this study guided students to reflect on their emotional and cognitive processes during the three phases of solving mathematics tasks:

1. Pre-learning forethought phase: pre-problem solving *emotional activities*, such as assessing tasks' easiness / difficulty.

2. During-learning phase: two central processes, monitoring and control. Monitoring concerns *emotional awareness* and control focuses on selecting and *adapting strategies* to manage the affect.

3. Post-learning phase: the learner's *affective reflections and reactions* after completing the problem.

Practically, the model modifies IMPROVE self-addressed questioning to increase students' awareness of their feelings ("How do I feel?"), guiding their reactions ("How shall I deal with negative / positive emotions?"), suggesting strategies (e.g. "Try to relax", "Take time out"), and reflections on the entire process ("How do I feel now?" and "Why?").

In a quasi-experimental study conducted in fifth grades, Tzohar-Rozen and Kramarski (in press) examined the effects of this programme on students' emotions towards learning mathematics. The study indicates that students in the experimental group exposed to RULER decreased their negative emotions and increased their self-efficacy more than those in the control group. In addition, the experimental group showed enhanced mathematical problem solving and transfer of knowledge. This result was supported by students' problem solving while thinking aloud and reflective interviews. Interestingly, even three months after the completion of the study the RULER students continued monitoring their emotions, using the affective-regulation strategy throughout the problem-solving phases more often than the control group. Table 6.2 presents quotes of students' responses by conditions.

The researchers concluded that the "affective regulation model" better equipped students to tackle negative feelings towards mathematics in addition to promoting the solution of CUN and transfer problems.

Table 6.2. **Quotes of students in RULER treatment and control groups**

Affective self-regulation group	Control group
Value of the programme: What do you remember about the programme?	
"We learned about feelings – positive and negative feelings – and how to cope with problem-solving situations. If I don't succeed I must tell myself that I can do it and that I must not give up...." "The most important thing is to check all the time so as not to despair and give up."	"We learned about math problems and all kinds of ways to solve them."
Programme's effectiveness: Did you find the strategy effective?	
"The strategy is very helpful. I still use it on difficult questions. Especially in math... If I don't feel like studying I take time out, relax a bit."	"I liked the questions." "I learned there are lots of patterns and how to discover the pattern."

Source: Tzhoar-Rozen, M. and B. Kramarski (in press), "How can an affective self-regulation program promote mathematical literacy in young students?", *Hellenic Journal of Psychology*.

Self-efficacy intervention with university students

Regarding adults, Hanlon and Yasemin (1999) investigated the effects of a metacognitive intervention designed to improve students' proficiency in mathematics. The intervention aimed to enhance students' self-efficacy through self-judgments of their mathematics performance. College-bound students participated in a five-week summer programme prior to their first year that included whole-class instruction, small group tutoring and individual meetings with instructional co-ordinators. As part of the intervention, the students made self-efficacy judgments regarding their ability to solve the problems on each of ten daily quizzes and compared these judgments to their maths quiz scores. In the individual meetings, the students identified short-term goals, created and maintained self-monitoring forms, and were introduced to maths heuristics. Over time, the students' achievement scores on a maths proficiency exam improved significantly, as did their confidence levels about passing the exam. Students who participated in the "self-efficacy" intervention group outperformed the control group who studied in the regular classes.

Social-emotional learning and its effects on social skills

Very few studies have focused on the effects of metacognitive interventions on social skills. One of the more comprehensive programs is that of Zins et al. (2004), called Collaboration for Academic, Social, and Emotional Learning (CASEL). CASEL's mission is to help make social and emotional learning an integral part of the education from preschool through high-school. The programme includes five steps: 1) recognising and managing emotions; 2) developing concern for others; 3) establishing positive relationships; 4) making responsible decisions; and

5) effectively handling challenging situations. These steps are intended to support students' positive behaviours and constructive social relationships, which in turn should foster academic learning.

Zins et al. (2004), presented findings based on a meta-analysis of 213 school-based social-emotional learning (SEL) programmes involving 270 034 children from kindergarten through to secondary school students. Compared to controls, the SEL participants showed significantly improved attitudes about the self and others, increased pro-social behaviour, increased motivation and school attendance, lower levels of problem behaviour and emotional distress, decreased anti-social behaviour within the class group, and increased academic performance, notably their achievement in mathematics and literacy (Greenberg et al., 2003; Zins et al., 2004). These findings show that metacognitive scaffolding can enhance social-emotional competencies among all ages.

Type III studies: the combined approach

Comparing the effects of different metacognitive pedagogies on motivation and self-efficacy

"What is better?" one may ask: to implement the modified version of metacognitive pedagogy that focuses on fostering motivation, or the original metacognitive pedagogy that aims to enhance cognitive outcomes, or maybe using a combined method which exposes students to both cognitive and social-emotional intervention? These questions might indirectly address the question of the extent to which the enhancement of socio-emotional competencies is a prerequisite for attaining cognitive-metacognitive goals, or by contrast, whether the enhancement of metacognitive skills is a prerequisite for fostering socio-emotional outcomes. These topics are important for both theoretical and practical reasons.

Michalsky (2013) addressed these questions in a recent study that investigated the effectiveness of cognitive-metacognitive versus motivational components. Michalsky implemented the study with tenth grade students during the reading of scientific texts. She utilised different versions of the IMPROVE self-addressed regulatory questions. Four research groups participated in this study. Three of them were exposed to one version or another of IMPROVE self-addressed questions: cognitive-metacognitive, motivational, or combined (cognitive-metacognitive and motivational). The fourth group received no self-addressed metacognitive questions and served as a control group. Table 6.3 shows the two kinds of self-addressed regulatory questions: cognitive-metacognitive and motivational. The dependent variables were scientific literacy and self-regulated learning (SRL) which were assessed off line by a questionnaire and online by the thinking aloud method. The self-regulated learning questionnaire includes the motivational and self-efficacy components.

Table 6.3. **Prompt types and self-regulated learning elements embedded in science reading comprehension texts**

Type of self-addressed question	SRL component	
	Cognitive-metacognitive	Motivational
Comprehension	What is the phenomenon all about? What is the problem/task needing investigation?	What makes you solve the problem/task? Explain. What will you do if you run into difficulties?
Connection	What do you already know about the phenomenon? What are the similarities/differences between the problem/task at hand and the problems/tasks you have encountered in the past? Please explain your reasoning.	What are the similarities/ differences between your efforts/ self-efficacy in the problem/task at hand and in the problems/ tasks you have solved in the past? Why?
Strategy	What are the inquiry strategies that are appropriate for solving the problem/task?	When/how should you implement a particular strategy to enhance your efforts to solve the problem/task? What "effort" strategies are appropriate for solving the problem/task?
Reflection	Does the solution make sense? Can you design the task in another way? How? Please explain your reasoning.	Do you feel good about your efforts/self-efficacy while comprehending the problem/ task? Can you motivate yourself in another way? How? Explain.

The findings indicate that all three treatment groups significantly outperformed the non-treatment control group on scientific literacy and on self-regulated learning behaviour, skills and beliefs. Of these, the fully combined cognitive-metacognitive-motivational support was most effective. Additional analyses showed no significant differences between the cognitive-metacognitive and motivational groups on scientific literacy (Figure 6.4). Michalsky (2013) explained that "mere exposure to reading scientific texts (the control group) is insufficient and that explicit instruction is required to train students to self-regulate their own learning" (Michalsky, 2013, p. 1864).

Michalsky further explains why the combined method resulted in higher levels of achievements and SRL than each of its components, and why no significant differences were found between the two components individually:

It seems that each of the three SRL components – cognitive, metacognitive, and motivation – is necessary but insufficient for self-regulation. For example, those who can regulate cognitive skills but are unmotivated to use them do not achieve the same level of performance as individuals who both possess the skills and are motivated to use them (Zimmerman, 2003). Similarly, those who are motivated but do not possess the necessary cognitive and metacognitive skills often fail to achieve high levels of self-regulation "…

As expected, the current findings further show the additional contribution of the combined cognitive-metacognitive-emotional approach beyond the contribution of each component by itself in enhancing science literacy and SRL" (p. 1986).

Figure 6.4. **Effect of cognitive, metacognitive and motivational interventions on scientific literacy**

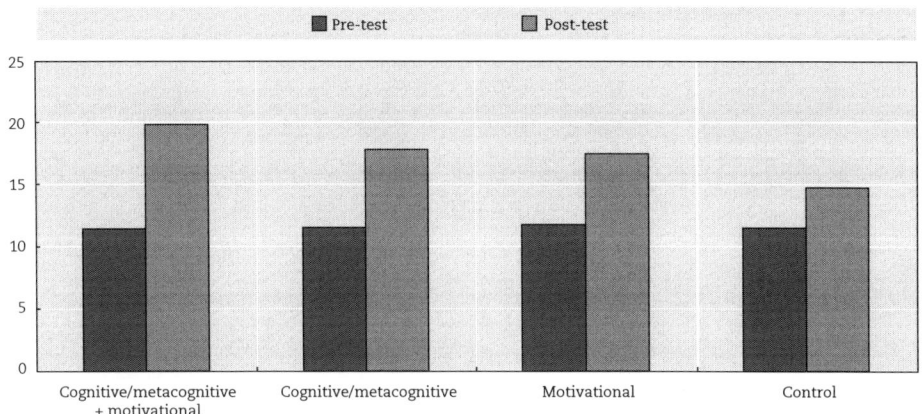

StatLink http://dx.doi.org/10.1787/888933149126

Source: Michalsky (2013), "Integrating skills and wills instruction in self-regulated science text reading for secondary students", *International Journal of Science Education*, Vol. 35(11), pp. 1846-1873.

Figure 6.5. **Effect of cognitive, metacognitive and motivational interventions on motivation**

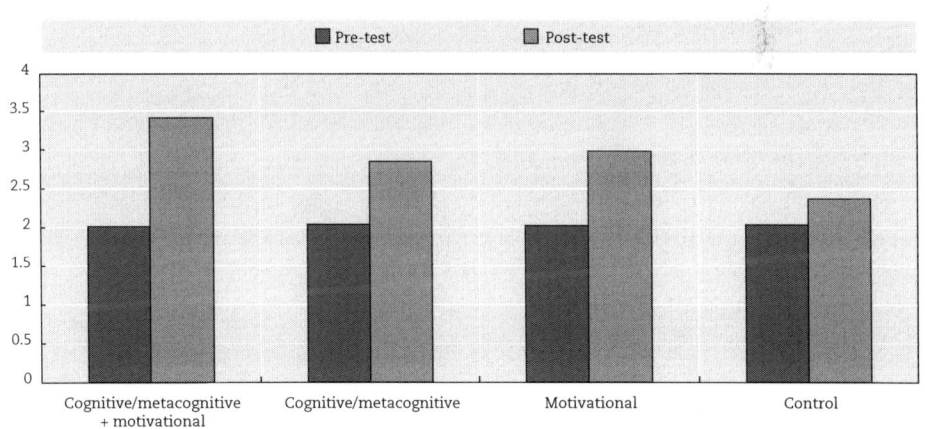

StatLink http://dx.doi.org/10.1787/888933149130

Source: Michalsky (2013), "Integrating skills and wills instruction in self-regulated science text reading for secondary students", *International Journal of Science Education*, Vol. 35(11), pp. 1846-1873.

Figure 6.6. **Effect of cognitive, metacognitive and motivational interventions on self-regulation**

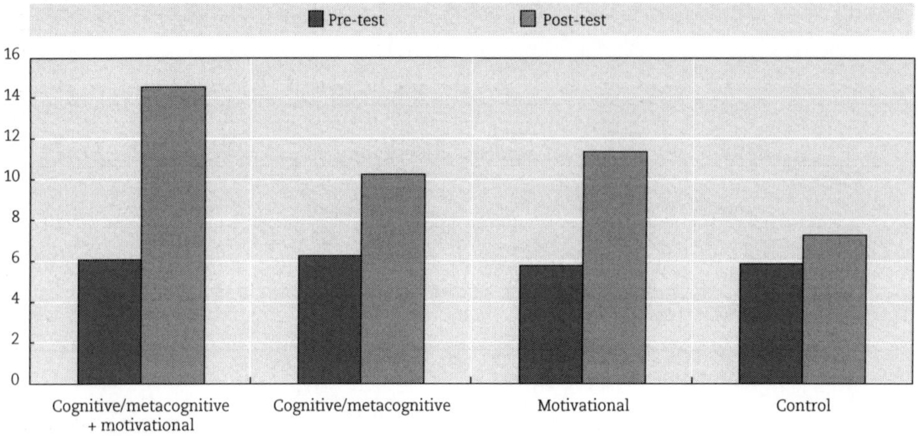

StatLink ⫘ http://dx.doi.org/10.1787/888933149144

Source: Michalsky (2013), "Integrating skills and wills instruction in self-regulated science text reading for secondary students", International Journal of Science Education, Vol. 35(11), pp. 1846-1873.

Teacher professional development

While Michalsky (2013) focused on high school students, Kohen and Kramarski (2012) investigated the effects of holistic self-regulation support (metacognition and motivation) on student-teachers' motivation, metacognition and academic outcomes. Participants (N = 97) were randomly assigned into one of two groups: *reflective support* in which a self-regulation scaffolding was provided, or *no support* (a control group).

The results indicate that the reflective support group displayed higher levels of metacognitive and motivational behaviour (interest and value, self-efficacy) and manifested less teaching anxiety than the group with no reflective support (for more detail on this study and how it was conducted, see Chapter 8). The findings are in line with previous studies showing the effects of the combined approach on cognitive and affective outcomes.

These two studies (Michalski, 2013 and Kohen and Kramarski, 2012) were implemented with adolescents (high school students) and adults (university students). The extent to which these findings are also applicable to younger students is still open. It is quite possible that the combined approach might have over-loading cognitive effects which would reduce its effectiveness with younger participants. This issue merits future research.

Conclusion

Social-emotional competencies are key factors in innovation-driven societies. The studies reviewed in this chapter show the conditions for promoting social-emotional competencies in addition to cognitive-metacognitive outcomes. The findings show that social-emotional skills do not need to be enhanced at the expense of cognitive and metacognitive outcomes. On the contrary, enhancing one kind of outcome (e.g. cognitive or emotional) often results in enhancing the other one as well.

The studies indicate that whether in kindergartens, primary schools, secondary schools and university, students who were exposed to IMPROVE or to other metacognitive pedagogies based on the same principles decreased their level of mathematics anxiety, increased their motivation and self-efficacy, and improved their social-emotional skills in addition to improving their academic achievement compared with a control group.

The findings also show that interventions that focus only on motivation or only on cognitive-metacognitive competencies are more effective than traditional instruction, but at the same time are less effective than the combined approach focusing on both motivation and metacognition. Another interesting finding is that no significant differences were found between students who were exposed to either one of these methods singly.

Given the importance of social-emotional competencies in innovation-driven societies, the findings presented in this chapter have important practical implications for teachers, educational policy makers, designers of effective learning environments, and further research:

- Since cognition and social-emotional competencies are intertwined in the brain, schools should be responsible for fostering social-emotional skills along with promoting cognitive outcomes.

- The use of explicit metacognitive pedagogies along with encouragement for students to articulate their thinking / emotions in small groups seem to be a desirable environment for enhancing both cognitive and social-emotional skills.

- Metacognitive pedagogies that were found to be effective for enhancing cognitive competencies were also effective for fostering social-emotional competencies. This finding is in line with research showing the importance of explicit metacognitive scaffolding not only with regard to cognition, but also for fostering social-emotional processes.

- Studies found that promoting student engagement by using metacognitive pedagogies was effective. Metacognitive pedagogies positively affect students' motivation to learn, reduce mathematics anxiety, enhance their self-efficacy, and actively involve them in the learning processes. However, all these studies might be affected by Hawthorne effects as the control groups were not exposed to another active treatment. This drawback should be addressed by further research.

- In all the studies reviewed in this chapter, the interventions were implemented by the classroom teachers. These findings indicate that social-emotional competencies similarly to metacognitive processes can be enhanced in ordinary schools.

- Finally, many issues are still open and merit future research. For example, little is known at present on the extent to which being engaged in CUN problems in itself affects social-emotional processes; who would benefit from it - younger or older students? Would the benefits for the lower achievers come at the expense of higher achiever? And under what conditions would the effects on social-emotional outcomes be optimal?

References

Baylor, A., E. Shen and D. Warren (2004), "Supporting learners with math anxiety: The impact of pedagogical agent emotional and motivational support", paper presented at the workshop on Social and Emotional Intelligence in Learning Environments, International Conference on Intelligent Tutoring Systems, Maceió, Brazil.

Boekaerts, M. (2010), "The crucial role of motivation and emotion in classroom learning", in N. Dumont, D. Istance, and F. Benavides (eds.), *The Nature of Learning: Using Research to Inspire Practice*, OECD Publishing, Paris, pp. 92-112, *http://dx.doi.org/10.1787/9789264086487-6-en*.

Brackett, M.A., S.E. Rivers, M.R. Reyes and P. Salovey (2012), "Enhancing academic performance and social and emotional competence with the RULER feeling words curriculum", *Learning and Individual Differences*, Vol. 22(2), pp. 218-224, *http://dx.doi.org/10.1016/j.lindif.2010.10.002*.

Bredekamp, S. and C. Copple (eds.) (1997), *Developmentally Appropriate Practice in Early Childhood Programs*, revised edition, NAYEC (National Association for the Education of Young Children), Washington, DC.

Crick, N.R. and K.A. Dodge (1994), "A review and reformulation of social information-processing mechanisms in children's social adjustment", *Psychological Bulletin*, Vol. 115(1), pp. 74-101.

Dobbs, D. (2006), "A revealing Reflection: Mirror neurons seem to affect everything from how we learn to speak to how we build culture", *Scientific American Mind*, May/June.

Efkelides, A. (2011), "Interactions of metacognition with motivation and affect in self-regulated learning: The MASRL model", *Educational Psychology*, Vol. 46(1), pp. 6-25.

Efklides, A. (2006), "Metacognition and affect: What can metacognitive experiences tell us about the learning process?", *Educational Research Review*, Vol. 1(1), pp. 3-14.

Gottman, J.M., LF. Katz and C. Hooven (1997), *Meta-Emotion: How Families Communicate Emotionally*, Lawrence Erlbaum, Mahwah, NJ.

Greenberg, M.T. et al. (2003), "Enhancing school-based prevention and youth development through coordinated social, emotional and academic learning", *American Psychologist*, Vol. 58(6-7), pp. 466-474.

Hanlon, E.H. and S. Yasemin (1999), "Improving math proficiency through self efficacy training", paper presented at the Annual Meeting of the American Educational Research Association, Montreal, Quebec, Canada, 19-23 April.

Hembree, R. (1990), "The nature, effects, and relief of mathematics anxiety", *Journal for Research in Mathematics Education*, Vol. 21(1), pp. 33-46.

Hinton, C. and K.W. Fischer (2010), "Learning from the development and biological perspective", in N. Dumont, D. Istance, and F. Benavides (eds.), *The Nature of Learning: Using Research to Inspire Practice*, OECD Publishing, Paris, pp. 113-134, *http://dx.doi.org/10.1787/9789264086487-7-en*.

Kohen, Z. and B. Kramarski (2012), "Developing self-regulation by using reflective support in a video-digital microteaching environment", *Education Research International*, Vol. 2012, Article ID 105246, *http://dx.doi.org/10.1155/2012/105246*.

Kramarski. B., I. Weiss and I. Kololshi-Minsker (2010), "How can self-regulated learning support the problem solving of third-grade students with mathematics anxiety?", *ZDM International Journal on Mathematics Education*, Vol. 42(2), pp. 179-193.

Mevarech, Z. R. and B. Kramarski (1997), "IMPROVE: A multidimensional method for teaching mathematics in heterogeneous classrooms", *American Educational Research Journal,* 34(2), pp. 365-395.

Mevarech, R.Z., T. Michalsky and C. Sasson (submitted), "Meta-cognition and science literacy: Immediate and transferred effects on science literacy, motivation, and self-esteem", paper submitted to the American Educational Research Association (AERA) annual meeting.

Michalsky, T. (2013), "Integrating skills and wills instruction in self-regulated science text reading for secondary students", *International Journal of Science Education,* Vol. 35(11), pp. 1846-1873.

Moreno, R., R.E. Mayer, H.A. Spires and J.C. Lester (2001), "The case for social agency in computer based teaching: Do students learn more deeply when they interact with animated pedagogical agents?", *Cognition and Instruction,* Vol. 19(2), pp. 177-213.

Pessoa, L. (2008), "On the relationship between emotion and cognition", *Nature Reviews: Neuroscience,* Vol. 9(2), pp. 148-158.

Pintrich, P.R. (2000), "Multiple goals, multiple pathways: The role of goal orientation in learning and achievement", *Journal of Educational Psychology,* Vol. 92(3), pp. 544-555.

Reynolds, C.R. and R.W. Kamphous (2004), *Behavior Assessment System for Children, Second Edition (BASC-2)* manual, North American Guidance Service, Circle Pines.

Reynolds, C.R. and R.W. Kamphous (1992), *BASC: Behavior assessment system for children,* North American Guidance Service, Circle Pines.

Richardson, F.C. and R.M. Suinn (1972), "The mathematics anxiety rating scale: Psychometric data", *Journal of Counselling Psychology,* Vol. 19(6), pp. 551-554.

Shen, E. (2009) "The effects of agent emotional support and cognitive motivational messages on math anxiety, learning, and motivation", PhD Thesis, Department of Educational Psychology and Learning Systems, Florida State University, *http://diginole.lib.fsu.edu/etd/309*.

Schraw, G., K.J. Crippen and K. Hartley (2006), "Promoting self-regulation in science education: Metacognition as part of a broader perspective on learning", *Research in Science Education,* Vol. 36, pp. 111-139.

Slavin, R.E. (2010), "Co-operative learning: What makes group-work work?, in N. Dumont, D. Istance and F. Benavides (eds.), *The Nature of Learning: Using Research to Inspire Practice,* OECD Publishing, Paris, pp. 161-178, *http://dx.doi. org/10.1787/9789264086487-9-en.*

Tzohar-Rozen, M, and B. Kramarski (in press), "How can an affective self-regulation program promote mathematical literacy in young students?", *Hellenic Journal of Psychology.*

Webster-Stratton, C. and M.J. Reid (2007), "Incredible years parents and teachers training series: A head start partnership to promote social competence and prevent conduct problems", in P. Tolin, J. Szapocznik, and S. Sambrano (eds.), *Preventing Youth Substance Abuse: Science-Based Programs for Children and Adolescents,* American Psychological Association, Washington, DC, pp. 67-88.

Webster-Stratton, C. and M.J. Reid (2004), "Strengthening social and emotional competence in young Children – The foundation for early school readiness and success: Incredible years classroom social skills and problem-solving curriculum", *Journal of Infants and Young Children,* Vol. 17(2), pp. 185-203.

Zeidner, M. (1998), *Test Anxiety: The State of the Art,* Plenum Press, New York.

Zins, J.E., M.R. Bloodworth, R.P. Weissberg and H.J. Walberg (2007), "The scientific base linking social and emotional learning to school", *Journal of Educational & Psychological Consultation,* Vol. 17 (2-3), pp. 191-210.

Zins, J.E., R.P. Weissberg, M.C. Wang and H.J. Walberg (2004), *Building Academic Success on Social and Emotional Learning: What Does the Research Say?,* Teachers College Press, New York.

Chapter 7

Combining technology and metacognitive processes to promote learning

Information and communications technology (ICT) could be a powerful tool for teaching mathematics, and particularly the solving of CUN tasks, but its potential has not always been fulfilled, possibly because 1) ICT-enhanced learning environments create cognitive overload; 2) meaningful learning with ICT depends on students being able to monitor, control and reflect on their learning; and 3) the type of metacognitive scaffolding provided in these environments needs to be tailored to the characteristics of the individual technologies. This chapter focuses on three kinds of ICT environments embedded within metacognitive pedagogies: specific maths software, general e-communication tools such as asynchronous learning networks, and general software such as e-books. Some of them are still in their infancy but they all appear to benefit from the addition of metacognitive scaffolding whether embedded into the technology itself or provided externally by a teacher.

The inclusion of ICT (information and communications technology) into mathematics education introduces additional challenges. Many of the current ICT environments are interactive, based on the assumption that learning is a constructivist process in which the learner has to be an active knowledge builder, while the computer's role is to facilitate this process (Mayer, 2010). ICT is particularly useful for approaching CUN problems because it enables the learners to search for information on the web, look for similar problems and sub-problems online, and use various computerised tools that can carry out the tedious work that is sometimes associated with solving mathematics problems (such as plotting graphs) and hence release cognitive energy for the higher-order cognitive processes (Mayer, 2010). Furthermore, search tools (such as Google) and online information sources, that are friendly, easy to use, and immediately responsive, may well lead users to reflect on the given information, decide which piece of information is most applicable to the given problem, what one can do with the information obtained, and whether or not the initial question should be modified in order to obtain a better response. Computer-supported collaborative learning, (e.g. asynchronous learning networks, forums or even emails) can become a powerful reflection tool, enabling students to be aware of how and why a solution path was chosen. The recorded learning interactions in such environments are often used for reflection purposes, planning, monitoring and control (e.g. Gama, 2004).

Unfortunately, however, the potential of these new technologies has not always been fulfilled. The very nature of ICT and the immediate feedback it provides often lead students to respond instantly, in a trial-and-error mode, without planning ahead, monitoring or controlling the solution processes. Consequently, many students in an ICT environment neither reflect on what they are doing nor generalise their performance for use in other contexts. Teachers, researchers, parents and even the students themselves complain sometimes that ICT leads learners to simply "press the buttons" instead of thinking about their thinking and performance: what they are doing and why they are doing it. Furthermore, the very fact that many ICT environments use visual, audio and moving stimuli simultaneously creates cognitive overload that further burdens the learning processes (e.g. Mayer, 2010). Hence, users need to be trained to apply monitoring, control and reflection processes in ICT environments (e.g. Azevedo and Hadwin, 2005). The new generation of ICT and their rapid spread raise the need for metacognitive scaffolding that can be integrated either within the software or implemented by the teacher to support learning.

ICT varies dramatically in terms of the kinds of software available. Mayer (2010) listed ten genres of technology-based learning environments:

1. computer-based training software usually designed in mastery-format in which the learner proceeds to the next section only after reaching a certain level of mastery;

2. multimedia;

3. interactive simulations where the learner can manipulate some parameters and observe what happens;

4. hypertext and hypermedia;

5. intelligent tutoring instructional systems that adjust what is presented according to the learner's knowledge;

6. inquiry-based information and information seeking tools such as Google;

7. animated pedagogical agents in which an on-screen character guides the learner through a computer-based lesson;

8. virtual environments;

9. games that serve instructional functions;

10. computer supported collaborative learning.

Lajorie (1993) organised computerised tools according to the cognitive functions the tools serve:

1. support cognitive and metacognitive processes;

2. share the cognitive load by providing support for lower-level cognitive skills so that the student may concentrate more on the higher level of cognitive activities;

3. allow learners to engage in cognitive activities that otherwise would be out of their reach, or construct the zone of proximal development in Vygotsky's terms (see Chapter 3);

4. provide learners the facilities to generate and test hypotheses in the context of problem solving.

In the area of mathematics, computerised tools can be classified into at least three broad categories: 1) specific maths software, such as a computer algebra system (CAS), graph plotter, or domain-specific software designed for specific purposes, usually remedial; 2) general e-communication tools, including distance learning, forums, asynchronous learning networks, or mobile learning; and 3) general software utilised for mathematics education, including intelligent cognitive tutor systems, hypertexts and e-books.

The sections below review some examples of ICT environments used in mathematics education for enhancing routine and CUN problem solving. It focuses mainly on metacognitive training implemented by teachers as supplements to the activities within the ICT environments.

Combining domain-specific mathematics software with metacognitive instruction

A computer algebra system (CAS) is an example of domain-specific software that provides online tools for manipulating mathematics expressions in symbolic forms. A CAS includes tools for simplifying expressions, plotting graphs, substituting variables, solving equations, carrying out algorithms, etc. CAS users believe that releasing students from doing the tedious work associated with mathematics manipulations enables learners to focus on the higher-order cognitive processes.

Kramarski and Hirsch (2003) tested this hypothesis by comparing students who used a CAS with or without the metacognitive scaffolding implemented via IMPROVE (Mevarech and Kramarski, 1997).

The study involved 43 eighth graders who practised 20 lab hours in total; about half of the students received metacognitive instruction, and the others did not (Kramarski and Hirsch, 2003). Students in the experimental group were asked to discuss the solution process in pairs by using the self-addressed questioning suggested by IMPROVE (Mevarech and Kramarski, 1997).

The results indicated that at the end of the study the CAS + IMPROVE group outperformed the CAS students only on algebraic reasoning and exploring patterns, but not on algebraic manipulations (solving equations and operations with algebraic expression) and analysing changes. In addition the CAS + IMPROVE group outperformed the CAS group on metacognitive knowledge of computerised learning environments. Qualitative analysis of protocols that recorded students thinking aloud during problem solving shows that during the discourse the IMPROVE students applied metacognitive statements more often (57%) than the no-IMPROVE students (38%). Figure 7.1 presents the mean scores of the different components of the mathematics post-test by learning conditions.

Figure 7.1. **Impact of IMPROVE on algebraic manipulation, reasoning, patterns and analysing changes**

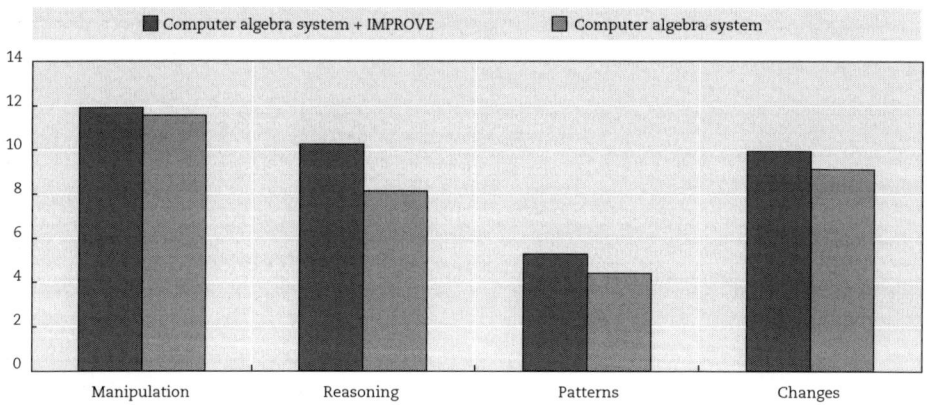

StatLink http://dx.doi.org/10.1787/888933149155

Source: Kramarski and Hirsch (2003), "Using computer algebra systems in mathematical classrooms", *Journal of Computer Assisted Learning*, Vol.19, pp. 35-45.

Jacobse and Harskamp (2009) took a similar approach, but one that better fits the nature of learning with ICT. In this study, conducted with fifth graders, the experimental group was allowed to *choose* metacognitive hints available on the web during mathematic problem solving; students in the control group studied "traditionally" with no computers and no metacognitive hints. Jacobse and Harskamp reported that the experimental group outscored the control group on the problem-solving post-test and they also improved their metacognitive skills. These results support the assumption that metacognitive skills can be enhanced by children having a free choice of metacognitive hints in a computerised learning environment and that the use of the hints can increase students' ability to solve word problems. However the researchers did not control for any Hawthorne effect that could obscure the positive effects caused by the metacognitive scaffolding.

In contrast, Gama (2004) constructed a metacognition instruction model called Reflective Assistant (RA) for solving algebra word problems in an interactive learning computer environment. RA focuses on three metacognitive skills: 1) problem understanding and knowledge monitoring; 2) planning and selecting metacognitive strategies; and 3) evaluating the learning experience. The RA software automatically builds a metacognitive profile of the student based on two measures. Knowledge-monitoring accuracy (KMA) measures the accuracy of the student's knowledge monitoring, and knowledge-monitoring bias (KMB) detects any systematic bias the students might exhibit in their knowledge monitoring. Box 7.1 provides examples of the RA reflective activities.

An empirical study conducted at the university level with undergraduate students exposed an experimental group to the Reflective Assistant and showed that students who performed the reflective activities spent more time on tasks, gave up on fewer problems and answered significantly more problems correctly than the control group. The evidence also suggested the RA model had positive effects on the students' metacognition.

Box 7.1. Examples of the Reflective Assistant's activities

1) Knowledge monitoring and performance

This activity focuses on comparing the solutions of previous problems to the current ones, asks students to judge their learning and shows them the accuracy of their judgments, and provides information on the time devoted to the solution of each problem. It also asks students to look for trends, check changes (improvements, stability, or withdrawal), and self-explain the reasons for the differences shown (if any exists).

2) Self-assessment of problem comprehension and difficulty

This activity aims to encourage students to reflect on their understanding and confidence that they can solve the problems correctly. It presents questions about the student's level of problem comprehension such as *Do you recognise the type of the problem? Do you understand the goals of the problem? Have you solved a similar problem before?*

3) Evaluation of problem-solving experience

This activity aims to give students an opportunity to review their most recent experience, explore why they acted as they did, what happened during problem solving, etc., as with the reflection-on-action proposed by Schoen (1987). The focus is on helping students to reflect on the "causes of their mistakes" with regard to the process, use of resources and time-management issues. In so doing students can develop a better understanding of their problem-solving experience and practice.

Source: Gama, C.A. (2004), "Integrating metacognition instruction in interactive learning environments", PhD Thesis, University of Sussex.

RA provides a graphical view of the student's problem-solving activities (Figure 7.2). This diagram was inspired by Schoenfeld's timeline graph originally created to analyse students' problem-solving behaviour (see Figure 4.2)

Figure 7.2. **Reflective activity for the evaluation of problem-solving experiences**

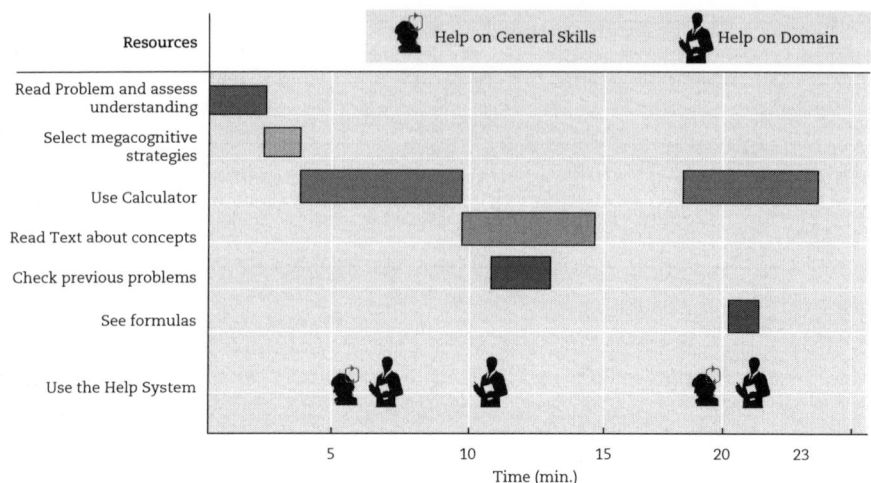

Source: Gama, C.A. (2004), "Integrating metacognition instruction in interactive learning environments", Ph.D. Dissertation, University of Sussex.

The diagram enables students to observe whether and when they used the main problem-solving resources available, the number of times the resources were used to solve a given problem, the time spent on reflective activities and the moments when the student asked for help. A textual explanation of the information depicted by the diagram is also available to trigger reflection on the student's performance (e.g. the student said the problem was difficult, but then she did not use any of the resources provided by the problem-solving environment to help her solving the problem, or she spent little time selecting strategies to solve the problem).

In all the studies reported above, the metacognitive intervention was either present or absent in the ICT environments. One may ask to what extent the effects of the metacognitive intervention depend on its features. Dresel and Haugwitz (2008) attempted to manipulate the quality of the metacognitive interventions provided by the computer by separating between the SRL training and the feedback. They assigned 151 sixth graders who studied mathematics with the aid of computers into one of three conditions: 1) students received the software that provided feedback and self-regulated training that focused on metacognitive control; 2) students received the software that provided only feedback with no metacognitive guidance; and 3) a control group that received neither. Results indicated that the self-regulation training led to better knowledge acquisition than feedback on its own, and that in both training conditions, students scored higher on motivation and knowledge acquisition than the control group.

An interesting series of studies implemented by Azevedo and his group examined the contribution of metacognitive scaffolding provided by a computer against that of a human facilitator. Azevedo and Jacobson (2008) compared the effect of self-regulated learning (SRL) with externally-facilitated SRL on adolescents' learning about the circular system while using hypermedia. Learners in the SRL group regulated their own learning, while learners in the externally regulated learning (ERL) group were exposed to a human tutor who facilitated their SRL. They found that learners under the ERL conditions gained statistically significantly more declarative knowledge and displayed a more advanced mental model on tests afterwards. Verbal protocols indicated that the ERL students regulated their learning by activating prior knowledge, engaged in several monitoring activities, deployed several effective strategies and engaged in adaptive help seeking. By contrast, learners under the SRL conditions used strategies ineffectively.

Taken together, these studies indicate that ICT environments based on specific maths programmes supported by metacognitive pedagogies enhance mathematics achievement, motivation and metacognitive processes more than using the same software with no metacognitive guidance. In addition, it was found to be more effective if a person provided the metacognitive scaffolding than if it was provided by the computer.

E-learning supported by metacognitive instruction

The term "e-learning" is a wide umbrella covering a range of settings in which students learn using technology, including online forums, emails and distance learning. In this section, we consider online forums (i.e. a site for posting messages to group members) implemented with or without metacognitive scaffolding.

In the area of mathematics, seventh grade learners who studied mathematics in online forums were compared to those learning in online forums supported by metacognitive scaffolding (Kramarski and Mizrachi, 2006). The findings indicated that the metacognitive group was better able than the other group to justify their reasoning, employed formal and logical arguments more often (90% versus 70%, respectively), and were more likely to explain how they obtained the answers rather than simply repeating the final solution (65.2% versus 30%, respectively). In addition, the online metacognitive group outperformed the other group in solving both textbook problems and an authentic task (the Pizza task, see Box 1.1); attained higher levels of mathematical reasoning, communicating with friends, using strategies, providing metacognitive feedback (e.g. monitoring, debugging and evaluation) and posing problems to given expressions; and were more motivated to solve mathematics problems in online discussion (Kramarski and Mizrachi, 2006; Kramarski and Dudai, 2009; Kramarski and Ritkof, 2002). Similar findings were also found in other studies integrating co-operative maths learning and metacognitive interventions into web-learning systems (e.g. Hurme, Palonen and Järvelä, 2006; Faggiano, Roselli and Plantamura, 2004).

Furthermore, experimental studies that compared learning science in forums that were either supported or not supported by metacognitive scaffolding showed without exception that the metacognitive support promotes learning in these environments (e.g. Azevedo, 2005; Azevedo and Jacobson, 2008; Azevedo et al., 2012).

An interesting question refers to the focus of the metacognitive scaffolding in co-operative settings with computer: should the scaffolding focus on the collaborative interaction (team regulation) or rather on the solution of the task (task regulation)? Saab, Joolingen and Hout-Wolters (2012) investigated this issue in science classrooms. Their study involved tenth-grade students who worked in pairs in a collaborative inquiry learning environment that was based on computer simulations (Sim Quest). Team regulation was provided by the RIDE rules: respect, intelligent collaboration, deciding together and encouraging (Table 7.1). Task regulation was provided through the Collaborative Hypothesis Tool (CHT) which used prompt windows to help students to formulate hypotheses together, plan experiments with the SimQuest programme and test their hypotheses on a scale from 0 to 100%. When the hypothesis was rejected, students were encouraged to follow the inquiry steps again until they confirmed the hypothesis. Students were randomly assigned into one of three conditions: 1) a group exposed to team regulation provided via RIDE (RIDE condition); 2) a group exposed to task regulation through the CHT as well as RIDE (CHT condition); and 3) a control group who were not exposed to any of these supports.

The results show that students overall used more team regulation than task regulation (see Figure 7.3). In both the RIDE condition and the CHT condition, students regulated their team activities more often than in the control condition. Moreover, in the CHT condition the regulation of team activities was positively related to the learning results. The authors concluded that different scaffolding via a simulation environment (SimQuest) affects team regulation differently, which in turn led to better learning outcomes.

Table 7.1. **RIDE rules and sub-rules taught through computerised instruction**

RIDE Rule	Sub-rules
(R) Respect	Everyone will have a chance to contribute
	Everyone's ideas will be thoroughly considered
(I) Intelligent collaboration	Sharing all relevant information and suggestions
	Clarify the information given
	Explain the answers given
	Give criticism
(D) Deciding together	Explicit and joint agreement will precede decisions and actions
	Accepting that the group (rather than an individual member) is responsible for decisions and actions
(E) Encouraging	Ask for explanations
	Ask till you understand
	Give positive feedback

Source: Based on SAAB et al. (2012).

Figure 7.3a **Mean scores for task regulation for RIDE and control groups**

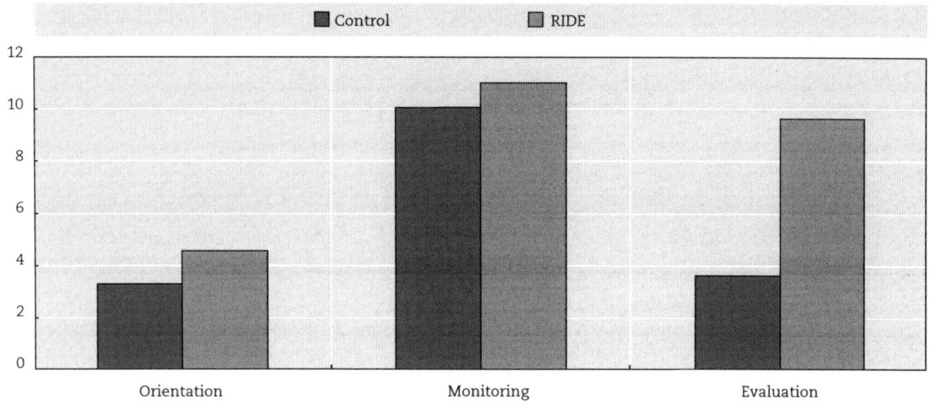

Note: RIDE corresponds to Respect, Intelligent collaboration, Deciding together, Encouraging.
Source: Saab, Van Joolingen and Van Hout-Wolters (2012), "Support of the collaborative inquiry learning process: influence of support on task and team regulation", *Metacognition and Learning*, Vol.7, Issue 1, pp. 7-23.

Figure 7.3b **Mean scores for team regulation for RIDE and control groups**

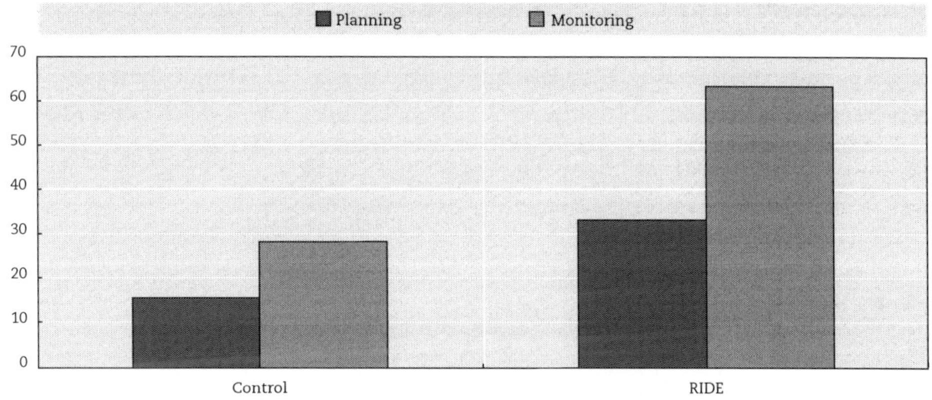

Note: RIDE corresponds to Respect, Intelligent collaboration, Deciding together, Encouraging.
Source: Saab, Van Joolingen and Van Hout-Wolters (2012), "Support of the collaborative inquiry learning process: influence of support on task and team regulation", *Metacognition and Learning*, Vol.7, Issue 1, pp. 7-23.

Asynchronous learning networks supported by metacognitive instruction

Asynchronous learning networks (ALN) refer to peer-to-peer online interactions to facilitate learning outside the constraints of time and place. ALNs have become quite popular within the educational arena. In higher education as well as in secondary schools, ALN courses are available in various disciplines, including mathematics and science. The benefits of ALNs are well known: learning can take place regardless of time and space; students from different schools can jointly perform inquiry projects or actively participate in solving problems; and students from different time zones can study co-operatively because the communication is asynchronous. However, the drawbacks of studying in ALN environments are similar to those found in co-operative face-to-face (F2F) settings: without explicit intervention, students may not be aware how to interact metacognitively. The following example shows how the embedding of metacognitive scaffolding within ALN improves achievement.

In a study conducted by Zion, Michalski, and Mevarech (2005), 407 tenth graders studied scientific inquiry for 12 weeks either through an asynchronous learning network or face-to-face, each with or without metacognitive instruction. Thus, the study compared four conditions: ALN with metacognitive guidance, ALN with no metacognitive guidance, F2F with metacognitive guidance and F2F without metacognitive guidance. All students were pre- and post-tested on scientific literacy (e.g. the ability to design experiments and draw conclusions) using a test designed by the authors, domain-specific inquiry skills in microbiology (the topic taught in all groups), metacognitive skills (assessed by the Metacognitive Awareness Inventory designed by Schraw and Davidson [1994]), and the quality of the students' discourse on the web. The findings indicated that the ALN with metacognitive group significantly outperformed the other three groups, while the F2F with no metacognitive support acquired the lowest mean score; no significant differences were found between F2F + metacognition and ALN with no metacognition. Further analyses (Mevarech, Zion and Michalsky, 2007) showed that the ALN + metacognition group outperformed the other groups on metacognition (both knowledge about cognition and regulation of cognition) and communication skills. These findings imply that ALN with metacognitive self-questioning is a promising learning environment, holding great potential for enhancing students' science literacy and metacognitive processes. Figure 7.4 presents the mean scores on science literacy by the four learning conditions.

Figure 7.4. **Impact of metacognitive guidance on scientific literacy in face-to-face or asynchronous learning environments**

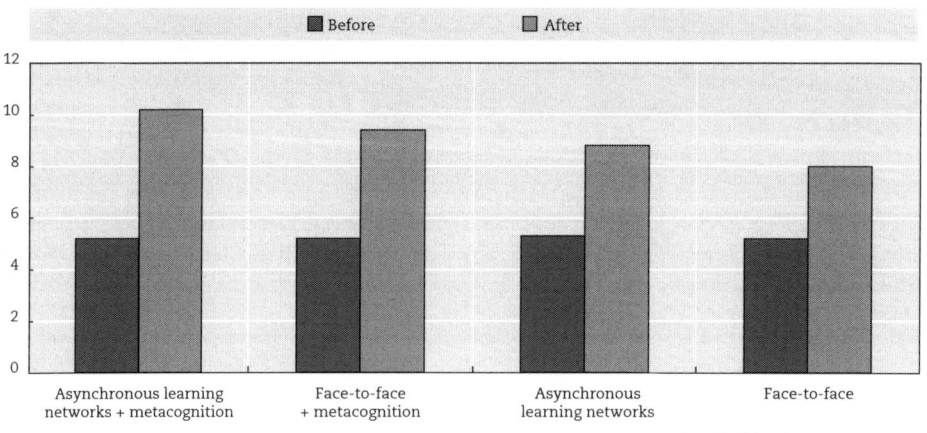

StatLink ⧉ http://dx.doi.org/10.1787/888933149186

Source: Zion, Michalsky & Mevarech (2005), "The effects of metacognitive instruction embedded within an asynchronous learning network on scientific inquiry skills", International Journal of Science Education, Vol. 27(8), pp. 957-958.

Mobile learning in mathematics

Mobile learning is a form of distance learning that uses cell-phone-based Short Message Service (SMS) text messages to offer tutoring that is not bound by time and space, and is characterised by short messages (about 140 characters). Advocates of SMS learning argue that the short messages lead the learner to focus on the main ideas delivered by the technology. These messages could deliver metacognitive prompts, such as "look for additional information / resources", "reflect on the solution process" or "evaluate the feasibility of the outcome". Those who oppose it use the same argument to claim that the short messages might be appropriate for enhancing recall, but not for facilitating complex problem solving that needs metacognitive guidance. For them, supporting SMS learning with metacognitive scaffolding contradicts the very nature of mobile learning (Stone, 2004a).

Although the use of SMS technologies in education is quite new, it has already been examined in several studies that reported inconclusive findings. Lu (2008) found that students who studied language by receiving SMS messages recognised more vocabulary than they did after receiving relatively detailed print material. Katz and Yablon (2011) report no significant differences in achievements between students who studied language via cell-phone based SMS messages and those who studied via email messages or received the material via conventional "snail mail" delivery. Yet Katz and Yablon (2011) found the students who received SMS messages were more motivated and felt that they had better control over their learning than the students who learned via email messages; both groups significantly outperformed the "snail-mail" group on these affective measures. The differences in the findings might result from the quality of the messages and interactions between students, but unfortunately, none of these studies examined these variables.

Hagos et al. (2009), Amiratashani (2010) and Stone (2004b) showed the effectiveness of mobile learning on mathematics outcomes, and especially on students' motivation, engagement and peer interactions. These studies also found that learners who learned by SMS supported by metacognitive scaffolding showed higher maths achievements compared to a control group who learned via SMS with no metacognitive guidance (e.g. Stone 2004).

Intelligent tutoring software

Mayer (2010) defines intelligent tutoring software (IST) as an "instructional system that tracks the knowledge of the learner and adjusts what is presented accordingly" (p. 181). The system provides problems and an individualised instruction based on the students' interactions with the programme. The issue of how to embed metacognitive scaffolds in such systems "has become crucial" (Azevedo and Hadwin, 2005, p. 367).

Roll, Aleven, McLaren and Koedinger (2007) provided a comprehensive review of ten principles regarding the integration of metacognitive scaffolding into an ITS for the teaching of geometry. They designed the Help Tutor – a metacognitive tutor to teach help seeking by students using the ITS. They investigated the additive effects of the Help Tutor software beyond the ITS. Roll et al. (2011) found that tenth and eleventh graders who used ITS with the Help Tutor were able to transfer their improved help-seeking skills to learning new domain-level content during the month following the intervention, when the support was no longer in effect.

Another direction was suggested by Aleven and Koedinger (2002) who investigated whether self-explanations can be effectivelyscaffolded using intelligent instructional software (Cognitive Tutor) in mathematics. They found that students who explained their steps during problem-solving practice with Cognitive Tutor gained better understanding of their learning and were more successful on transfer problems compared with students who did not explain their steps. The authors interpreted these results: "By engaging in explanation, students acquired better-integrated visual and verbal declarative knowledge and acquired less shallow procedural knowledge" (Aleven and Koedinger, 2002, p. 147). These findings are expected since also in non-ICT environments the explainers benefit more than those who receive the explanations (Webb, 2008).

Mathematics e-books

In contrast to printed books, e-books are interactive books that include visual, audio, and moving stimuli. E-books can "read" aloud the text or present moving objects, and allow readers to enlarge or change fonts, search for key terms online, or highlight, bookmark and annotate content. E-books are being rapidly introduced into schools and kindergartens for reading as well as for teaching, including the teaching of mathematics (Shamir and Baruch, 2012).

Just as with printed materials, learning with e-books varies tremendously and often needs to be supported by metacognitive scaffolding in one form or another. Such metacognitive prompts could be an integral part of the e-book, or be provided by the teacher during classes with the e-books.

Recently, Shamir and Baruch (2012) designed a mathematics e-book for kindergarten children who are at risk of developing a learning disability. The e-book contains a mathematics story, pictures, moving images, mathematics games, dictionary, and a "reader" who "reads" the story loudly. Children can look for links that provide explanations of difficult words, concepts, and so on. They can also ask for metacognitive prompts.

Shamir and Baruch administered the mathematics e-book to two groups of kindergarten children: one group received the e-book with metacognitive prompts and the other group used the e-book with no metacognitive prompts. A third group served as a control group without being exposed to either the e-book or the metacognitive prompts. Results indicated that while both groups who used the e-book improved their mathematical knowledge more than the group who studied the same mathematics concepts with no e-book, no significant differences were found between children using the e-book with or without the metacognitive prompts. Unfortunately, the authors did not collect data on the extent to which the children asked for the metacognitive prompts and how they used them (if at all). Thus, it is difficult to interpret the findings. In any case, the e-book technology is only in its infancy and it is too early to draw any further conclusions.

Conclusion

While the new-generation technologies have great potential to promote CUN solutions and certain mathematics skills, they have not created a revolution in mathematics education, probably because students and teachers seem to have difficulties in monitoring, controlling and reflecting on the learning processes while using them. The good news is that this could be easily changed by embedding metacognitive scaffolding into ICT environments, either by integrating the metacognitive prompts into the software or having it presented by the teacher. Generally speaking, the research suggests that ICT environments supported by metacognitive scaffolding enhance students' mathematics reasoning, mathematics communication, CUN problem solving, science literacy, motivation and self-efficacy. These last two variables (motivation and self-efficacy) are important outcomes by themselves.

However, discussing the effects of "ICT" on learning is too vague and too broad. It is like asking what effects books have on schooling outcomes. Obviously, it depends on the book itself, the reading process, the learner and the expected outcomes. Similarly, learning with mathematics software such as CAS or remedial maths software is different from learning it with an asynchronous network, e-book, SMS, Google, computer games, intelligent cognitive system or Wikipedia, to mention but a few. The two general evidence-based conclusions that we can draw at this point are: 1) the effectiveness of ICT environments depends by and large not only on the ICT characteristics, but also on students' abilities to monitor, control, and reflect on the learning process and the outcomes; and 2) metacognitive scaffolding can be modified to suit the distinctive characteristics of the different types of ICT, as described below.

- In using maths remedial software, the role of the technology is to present appropriate exercises (e.g. exercises that fit the learner's capability), and provide feedback (rewards or punishments) according to the learner's response. The role of the metacognitive scaffolding is to guide learners to analyse the similarities and differences between the exercise on the screen and those solved in the past, plan strategies for implementation, reflect on the feedback, judge their learning, look backward, and decide what resources are still needed in order to attain mastery.

- In ICT environments that provide tools to support the lower-level cognitive skills, the metacognitive scaffolding guides student to concentrate on the higher level cognitive activities. For example, CAS simplifies mathematics expressions and thereby enables students to focus on the core of the problem, without being bothered by the tedious work of simplifying expressions. Similarly, graph plotters construct the graphs and thereby free up students' cognitive resources for analysing and enquiring on the functions themselves. In these cases, the metacognitive scaffolding guides students to think what the "whole" problem is all about, what they should do with the resulting expressions/graphs etc., and how the technology could help them in solving the CUN task.

- Special ICT devices (e.g. dynamic geometry environments) have been developed to allow learners to generate and test hypotheses in the context of problem solving. The metacognitive scaffolding is particularly beneficial in these open environments where the learners can become lost in the openness and the inherent complexity of the materials. In these environments, the metacognitive scaffolding trains students to make hypotheses, plan how to test the hypotheses, and think how to organise the information obtained for testing the hypotheses. Reflecting on the outcomes allows the learner to generalise the outcome, or alternatively continue the investigation, sometimes by rephrasing the original hypotheses.

- In ICT environments that support information acquisition (e.g. Wikipedia, online maths webs), the role of the technology is to provide the links, whereas the role of the metacognitive scaffolding is to guide learners in organising the information, relating it to previous knowledge, judging their comprehension and planning resource allocation accordingly.

- Learning with SMS frequently aims at improving rote memory of facts (e.g. recall a formula); metacognitive scaffolding could be integrated into the SMS system.

- Learning with multimedia, hypertexts, interactive simulations, e-books, etc., is often associated with cognitive overload caused by the rich visual, auditory and moving stimuli that are presented on the screen. The metacognitive scaffolding should aim to reduce cognitive overload by guiding learners to focus on the essential topic and ignore the extraneous stimuli, and by training the learner to manage the learning processes, organise the information, and integrate it with previous knowledge.

- In ICT environments that include texts, the metacognitive scaffolding should also focus on reading comprehension by leading the users to implement reading strategies such as underlining important words or concepts, drawing flowcharts, explaining difficult words, or summarising the text.

- In asynchronous learning networks, an interactive learning environment that provides ample opportunities for students to interact with other students, the metacognitive scaffolding should guide students to articulate and formulate their thinking, reflect on their own and other's ideas, and communicate mathematically.

In summary, many of these new-generation ICT environments aim not only to facilitate information acquisition, but mainly and more importantly to enhance learners' knowledge construction. In these environments, learners must activate metacognitive processes in order to build their knowledge, while the computer facilitates the learning processes by various means (e.g. interactive simulation, multimedia, hypertext, intelligent tutoring systems, computer-supported collaborative learning, or search tools). The metacognitive scaffolding facilitates the monitoring, control and reflection processes, in addition to providing more domain-specific metacognitive prompts that lead learners to focus on the higher order cognitive levels, pay attention and select the relevant information, organise the information obtained, integrate and manipulate it for further use, formulate and test hypotheses, and plan how to use the specific device for solving routine and CUN tasks.

References

Aleven, V.A.W.M.M. and K.R. Koedinger (2002), "An effective metacognitive strategy: Learning by doing and explaining with a computer-based Cognitive Tutor", *Cognitive Science*, Vol. 26, pp. 147-179.

Amiratashani, Z. (2010), "M-learning in high school: The impact of using SMS in mathematics education – an Iranian experience", Fourth International Conference on Distance Learning and Education (ICDLE), San Juan, Puerto Rico, 3-5 October.

Azevedo, R. (2005), "Using hypermedia as a metacognitive tool for enhancing student learning? The role of self-regulated learning", *Educational Psychologist*, Vol. 40(4), pp. 199-209.

Azevedo, R. and A.F. Hadwin (2005), "Scaffolding self-regulated learning and metacognition: Implications for the design of computer-based scaffolds", *Instructional Science*, Vol. 33, pp. 367-379.

Azevedo, R. and M.J. Jacobson (2008), "Advances in scaffolding learning with hypertext and hypermedia: A summary and critical analysis", *Educational Technology Research and Development*, Vol. 56(1), pp. 93-100.

Azevedo et al. (2012), "Using artificial pedagogical agents to examine the role of metacognitive processes during learning with MetaTutor", in *Metacognition 2012 – Proceedings of the 5th Biennial Meeting of the EARLI Special Interest Group 16 Metacognition*, Milan, 5-8 September.

Dresel, M. and M. Haugwitz (2008), "A computer-based approach to fostering motivation and self-regulated learning", *The Journal of Experimental Education*, Vol. 77(1), pp. 3-18.

Faggiano, E., T. Roselli and V.L. Plantamura (2004), "Networking technologies to foster mathematical metacognitive processes", in *Proceeding of the 4th IEEE International Conference on Advanced Learning Technologies*, IEEE, California.

Gama, C.A. (2004), "Integrating metacognition instruction in interactive learning environments", PhD Thesis, University of Sussex, *http://homes.dcc.ufba.br/~claudiag/thesis/Thesis_Gama.pdf*.

Hagos, L.C. (2008), "Enhancing teaching and learning through SMS-mediated learning in mathematics", in J. Fong, R. Kwan and F. L. Wang (eds.), *Hybrid Leaning: A New Frontier*, City University of Hong Kong, Hong Kong, China, pp. 33-42.

Hurme, T-R., T. Palonen and S. Järvelä (2006), "Metacognition in joint discussions: An analysis of the patterns of interaction and the metacognitive content of the networked discussions in mathematics", *Metacognition and Learning*, Vol. 1(1), pp. 181-200.

Jacobse, A.E. and E.G. Harskamp (2009), "Student-controlled metacognitive training for solving word problems in primary school mathematics", *Educational Research and Evaluation*, Vol. 15(5), pp. 447-463.

Katz, Y.J. and Y.B. Yablon (2011), "Affect and digital learning at the university level", *Campus-Wide Information Systems*, Vol. 28(2), pp.114-123.

Kramarski, B. and V. Dudai (2009), "Group-metacognitive support for online inquiry in mathematics with differential self-questioning", *Journal of Educational Computing Research*, Vol. 40(4), pp. 365-392.

Kramarski, B. and C. Hirsch (2003), "Using computer algebra systems in mathematical classrooms", *Journal of Computer Assisted Learning*, Vol. 19, pp. 35-45.

Kramarski, B. and N. Mizrachi (2006), "Online discussion and self-regulated learning: Effects of instructional methods on mathematical literacy", *The Journal of Educational Research*, Vol. 99(4), pp. 218-230.

Kramarski, B. and R. Ritkof (2002), "The effects of metacognition and email interactions on learning graphing", *Journal of Computer Assisted Learning*, Vol. 18(1), pp. 33-43.

Lajoie, S. P. (1993), "Computer environments as cognitive tools for enhancing learning", in S.P. Lajoie and S.J. Derry (eds.), *Computers as Cognitive Tools*, Routledge, NY, pp. 261-288.

Lu, M. (2008), "Effectiveness of vocabulary learning via mobile phone", *Journal of Computer Assisted Learning*, Vol. 24(6), pp. 515-525.

Mayer, R.E. (2010), "Learning with technology", in H. Dumont, D. Istance and F. Benavides (eds.), *The Nature of Learning: Using Research to Inspire Practice*, pp. 179-195, OECD Publishing, Paris, pp. 179-195, http://dx.doi.org/10.1787/9789264086487-10-en.

Mevarech, Z. R. and B. Kramarski (1997), "IMPROVE: A multidimensional method for teaching mathematics in heterogeneous classrooms", *American Educational Research Journal*, 34(2), pp. 365-395.

Mevarech, Z.R., M. Zion and T. Michalsky (2007), "Peer-assisted learning via face-to-face or a-synchronic learning network embedded with or without metacognitive guidance: The effects on higher and lower achieving students", *Journal of Cognitive Education and Psychology*, Vol. 4(3), pp. 456-471.

Roll, I., V. Aleven, B. McLaren and K.R. Koedinger (2007), "Designing for metacognition: Applying cognitive tutor principles to the tutoring of help seeking", *Metacognition and Learning*, Vol. 2(2-3), pp. 125-140.

Saab, N., W.R. Joolingen and B.H.A.M. Hout-Wolters (2012), "Support of the collaborative inquiry learning process: Influence of support on task and team regulation", *Metacognition and Learning*, Vol. 7(1), pp. 7-23.

Schoen, D.A. (1987), "Teaching artistry through reflection-in-action", in *Educating the Reflective Practitioner,* Jossey-Bass Publishers, San Francisco, CA, pp. 22-40.

Shamir, A. and D. Baruch (2012), "Educational e-books: A support for vocabulary and early mathematics of children at risk for learning disabilities", *Educational Media International,* Vol. 49(1), pp. 33-47.

Stone, A (2004a), "Mobile scaffolding: An experiment in using SMS text messaging to support first uear university students", in *Proceedings, IEEE International Conference on Advanced Learning Technologies,* IEEE, pp. 405-409.

Stone, A. (2004b), "Designing scalable, effective mobile learning for multiple technologies", in J. Attewell and C. Savill-Smith (eds.), *Learning with Mobile Devices,* Learning and Skills Development Agency, London.

Webb, N.M. (2008), "Learning in small groups", in T.L. Good (ed.), *21st Century Education: A Reference Handbook,* SAGE Publications, Los Angeles, pp. 203-211.

Zion, M., T. Michalsky and Z.R. Mevarech (2005), "The effects of metacognitive instruction embedded within an asynchronous learning network on scientific inquiry skills", *International Journal of Science Education,* Vol. 27(8), pp. 957-958.

Chapter 8

Metacognitive programmes for teacher training

Teachers and principals have an important role in introducing change in schools. Given that "one cannot teach what one does not know", teachers' own metacognitive skills are increasingly being studied. Observations have shown that although teachers seldom explicitly activate metacognitive processes while teaching, they do apply them implicitly in the classroom. Their understanding of metacognition is related not only to their practice, but also to their students' self-regulated learning and achievement. Professional development programmes are the natural settings for the introduction of innovative teaching methods. Studies into the effects of metacognitive pedagogies on both in-service and pre-service teachers have found they positively enhanced teachers' knowledge, pedagogical-content knowledge (PCK), self-regulated learning (SRL) and self-efficacy, but these studies have not followed teachers into the classroom.

Studies have shown that teachers in general (Perry, Phillips and Hutchinson, 2006; Zohar, 1999), and mathematics teachers in particular (Verschaffel et al., 2007), seldom explicitly activate metacognitive processes during teaching (Putnam and Borko, 2000; Randi and Corno, 2000). Furthermore, both in-service and prospective math teachers do not monitor their comprehension effectively, use correct representations of the problem conditions, or properly apply other metacognitive skills during their own solution of non-routine problems (Nool, 2012; Koren, 2008). It is crucial, therefore, to acquaint teachers with the principles of metacognition.

The research in this field has addressed three basic issues, all related to the impact of metacognition on teachers' work:

1. How do teachers apply metacognitive processes in their classrooms, and what are the relationships between teachers' and students' metacognition?

2. How can teachers' metacognition be enhanced via professional development programmes?

3. What are the effects of metacognitive pedagogies on pre-service teachers?

In contrast to previous chapters that focused mainly on studies of mathematics or science education, this chapter also includes studies in which the participants were general teachers, who teach all subjects with no distinction between disciplines.

How do teachers apply metacognitive processes in their classrooms?

Assuming that enhancing students' metacognition is essential for promoting learning, and that "one cannot teach what one does not know", teachers' metacognition has become an important research area. Studies have used observations, thinking aloud, lesson plans, structured interviews and questionnaires to assess teachers' metacognitive behaviours either in "real time" during teaching or off-task, when teachers are asked to reflect back on their activities. In particular, researchers have asked what kind of metacognitive behaviour teachers activate in their classrooms and how it affects students' self-regulated learning (SRL) and achievement.

In 1998, Artzt and Armour-Thomas observed, analysed videotapes of, and interviewed seven expert and seven novice mathematics teachers in order to identify the kinds of metacognitive activities that teachers implement in maths classrooms. They showed that teachers activate metacognitive processes during lesson planning by sequencing the tasks according to previous student understanding, interest, and curiosity. During the lessons, teachers monitor and regulate teaching by adapting instruction based on the information received through monitoring students learning and interest. For example, teachers added examples to increase student understanding or excluded examples in order to save time. Finally, metacognitive-oriented teachers assessed their accomplishment of their goals in terms of students' understanding and content coverage. They also revised instruction as needed.

However, Artzt and Armour-Thomas did not implement any interventions, nor did they report the extent to which there were significant differences between experts and novice teachers in metacognition. More importantly, these researchers did not examine the relationships between teachers' and students' metacognition.

About a decade later, Wilson and Bai (2010) took the research on teachers' metacognition one step further by analysing teachers' pedagogical understanding of what metacognition is and how it is related to teaching students to be metacognitive. In this study, the pedagogical understanding of metacognition involved understanding the nature of what it means to teach and how students learn strategies that encourage them to be metacognitive. Data analysis of 105 teachers from kindergarten through to upper secondary schools indicated that the participants' metacognitive knowledge had a significant impact on their pedagogical understanding of metacognition. Wilson and Bai found that teachers' understanding of metacognition appear to be related to their perceptions of the instructional strategies that assist students in becoming metacognitive. The results further revealed that "teachers who have a rich understanding of metacognition report that teaching students to be metacognitive requires a complex understanding of both the concept of metacognition and metacognitive strategies" (Wilson and Bai, 2010, p. 269).

Wilson and Bai (2010) further demonstrated that an individual teacher's understanding of metacognition was related to the instructional strategies they perceived to be effective in helping students to become metacognitive. Those strategies include: demonstration, scaffolding, teaching conditional knowledge, providing students with appropriate assignments that assist their metacognitive thinking, and taking the time to help students to be self-aware of cognitive processes. According to the teachers' understanding of metacognition, a metacognitive person is someone who monitors his/her understanding and uses strategies to regulate understanding.

On the basis of this study, Wilson and Bai (2010) concluded that teachers may benefit from professional development programmes that emphasise the differences between engagement and awareness when guiding students to implement metacognitive strategies. In order to assist pre-service and practising teachers in developing this complex pedagogical understanding of metacognition, teacher educators should focus on declarative, procedural and conditional metacognitive knowledge and how these three components relate to the application of metacognition. As Veenman et al. put it, "Teachers are absolutely willing to invest effort in the instruction of metacognition within their lessons, but they need the tools for implementing metacognition as an integral part of their lessons." (Veenman et al., 2006, p. 10, quoted in Wilson and Bai, 2010).

This conclusion raises the issue of the kind of "tools" and environments that could affect teachers' metacognition. Prytula (2012) addressed this issue with regard to professional learning communities (PLCs). He reports that PLCs played an effective role in allowing teachers to reflect on their own and others' metacognition because they give teachers ample opportunities to articulate their thinking, discuss problems raised during teaching and examine their own beliefs. Prytula further showed that each participant used their understanding of their own metacognition to influence the learning of others. According to Prytula, for teachers, being aware of their metacognition is "a crucial component in teaching and learning, where the teacher is not only able to reflect on his/her thinking for the purpose of solving problems, but is actually aware of the type of thinking strategies that are being used in certain environments" (Prytula, 2012, p. 113). Prytula concluded that pre- and in-service professional development needs to shift from mastery of skills to metacognition.

Implementing metacognitive pedagogies in professional development programmes

Professional development programmes are the natural place to introduce metacognitive pedagogies to teachers. There are at least two ways to do so: the first is by delivering theoretical knowledge (e.g. lecturing and raising awareness), and the second by exposing teachers to metacognitive pedagogies followed by actual application of the method. Given the large body of research showing the greater effectiveness of "learning by doing" or "learning from experience" over theoretical training (e.g. Kolb, 1984), this section reviews the results of four studies based on the second approach that emphasises "learning by doing". These studies were implemented either in "traditional" environments that focused on domain-specific knowledge, such as maths and science, or in ICT-based professional development programmes in which the main objective was to facilitate the use of various kinds of technologies. They covered professional development courses in geometry (Koren, 2008), arithmetic (Kramarski and Revach, 2009; Mevarech and Shabtay, 2012), and ICT (e.g. Phelps et al., 2004).

Teachers' creativity

Koren (2008) was one of the first researchers to examine the effects of a metacognitive pedagogy (IMPROVE) on primary school teachers participating in a professional development course that focused on geometry. Koren chose to study the teaching of geometry with or without metacognitive scaffolding because teachers face considerable difficulties in teaching this subject at all levels of education (e.g. Swafford, Jones and Thronton, 1997).

The participants were 30 primary school maths teachers, recruited from 17 schools (Koren, 2008). Teachers were randomly assigned into one of two groups: half of them were exposed to IMPROVE and the others studied the same topics via traditional instruction with no explicit metacognitive training. The topic they studied was "polygons: perimeters and areas" and the studied outcomes were the teachers' pedagogical content knowledge (PCK) and achievement in geometry. The teachers' PCK (Shulman, 1986) was assessed on four criteria:

1. Appropriateness: the extent to which the provided explanations fit the pedagogical situation (scores ranged from 0-3).

2. Terminology: the extent to which the teacher used the correct maths terminology in her explanations (scores ranged from 0-2).

3. Creativity: the extent to which the teacher applied original ideas during the mathematics discourse with the students (scores ranged from 1 -3).

4. A general PCK evaluation (scores ranged from 1 [very poor] to 6 [excellent]).

In addition, teachers were tested before and after the intervention on achievement in geometry.

Although no significant differences were found between the groups on their post-test achievement in geometry, significant differences were found between the groups on their geometry PCK (Effect Size = .30) and creativity (Effect Size = .88) (Figure 8.1). The IMPROVE teachers were better able to provide original ideas in their explanations than the control group.

Figure 8.1. **Impact of IMPROVE on teachers' pedagogical content knowledge**

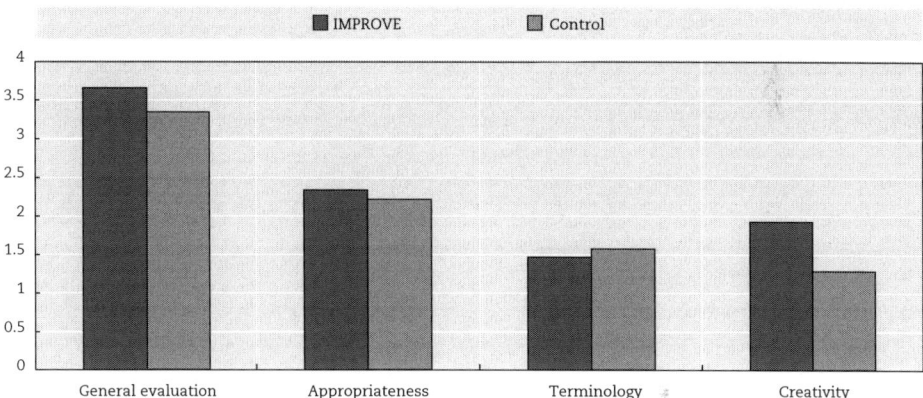

StatLink ⟐ http://dx.doi.org/10.1787/888933149192

Source: Koren (2008), "The effects of metacognitive guidance on teachers' mathematics content knowledge and pedagogical knowledge: The case of polygons' areas and perimeters", MA Thesis, Bar-Ilan University (Hebrew).

Three factors may explain these findings. First, it is possible that significant differences were only obtained for creativity because the IMPROVE teachers frequently addressed the "reflection" self-addressed question that guides them to think of the problems in different ways. Thus, IMPROVE teachers were trained to reflect on the problem-solving process by suggesting additional solutions, even when the response they had obtained was correct. Second, the duration of the study (four weeks) was too short to cause significant differences between the groups on achievement and other PCK components. Finally, while the teachers in this study experienced the application of the metacognitive processes when they themselves solved the geometry problems, they did not practice the implementation of the metacognitive pedagogy in their classrooms. One may argue that such "learning from experience" is not sufficient to introduce deep change in teachers' metacognition and only "learning by doing" would be effective. This hypothesis merits future research.

Teachers' pedagogical content knowledge (PCK)

Kramarski and Revach (2009) considered professional development programmes from a different perspective. They assumed that professional development settings turn participants into teachers and learners at the same time. Thus, they argued that the four self-addressed metacognitive questions used by IMPROVE (Mevarech and Kramarski, 1997) have to be modified to include both perspectives: that of the learner (i.e. mainly referring to problem solving) and that of the teacher (i.e. mainly referring to planning, monitoring and elaboration of the teaching process). Table 8.1 presents the resulting modified metacognitive self-addressed questions.

Table 8.1. **Modifying IMPROVE for teacher and student training**

IMPROVE questioning	Learner's perspective	Teacher's perspective
Comprehension questions: Structure of the task	What is the problem about? Identify: type of problem mathematical terms the givens the question	What is the goal or main idea of the lesson? Demonstrate: the lesson's topic mathematical knowledge explanations needed in the lesson
Connection questions:	What is the similarity or the difference between the two problems/explanations? WHY? Write down your reasons.	What is the similarity or the difference between the two lessons/examples? WHY? Write down your reasons.
Strategic questions: declarative (what), procedural (how), conditional (why)	What strategy/tactic/principle can be used and how in order to solve the problem/task? WHY? Write down your reasons.	What strategy/tactic/principle can be used and how in planning/teaching the lesson? WHY? Write down your reasons.
Reflection questions: Monitoring and evaluation - during and after the process	Do I understand? Is the solution reasonable? What is a good mathematical argument? Can I solve the task differently?	Which difficulties am I expecting in the lesson? How can I achieve my goals in the lesson? What is a good mathematical argument? Are the students engaged in the lesson? Can I plan the task differently?

The participants in the Kramarski and Revach (2009) study were 64 elementary school maths teachers who participated in a month-long professional development programme aimed at enhancing their mathematical and pedagogical knowledge. The study was part of a three-year professional development programme sponsored by the Israeli Ministry of Education. Teachers were randomly assigned into one of two groups: those who were exposed to the modified version of IMPROVE as described above, and a control group with no metacognitive intervention. The outcomes assessed included

PCK based on the analysis of teachers' lesson plans and teachers' mathematical knowledge. The lesson plans were assessed on three categories: task demands, task design and teaching approach. On each category scores ranged from 0 to 3, thus giving a total score ranging from 0-9. Teachers' mathematical knowledge was tested before and after the intervention on seven items; scores ranged from 0 to 7. The pre-test was based on the "apple tree task" selected from PISA 2003 (p. 96-97), and the post-test presented an authentic task plotting a graph showing the patterns of money saving by two children. In addition, the study reported the findings on a delayed end-of-year teacher test administered annually by Israeli Ministry of Education. That test covered a large range of mathematical and pedagogical knowledge. For the sake of simplicity all scores were transformed into percent correct items.

Figure 8.3 shows the mean scores for mathematical knowledge and Figure 8.4 the scores for PCK by time and condition. The graphs show that the IMPROVE teachers outperformed the control group on the total score of mathematical knowledge (Mean = 88.6 and 76.8; Standard Deviation = 7.56 and 10.8, for the IMPROVE and control groups, respectively; $F(1,61) = 14.25$, $p<.0001$), and on the total score for planning a lesson $(F(3,60) = 19.17$; $p<.0001)$ as well all of its components except task design. The follow-up test administered by the Ministry of Education also indicated significant differences between the groups on both mathematical knowledge (Mean = 87.7 and 75.66; Standard Deviation = 7.56 and 10.11 for IMPROVE and control groups, respectively; $F(1,62) = 30.30$, $p<.001$) and pedagogical knowledge (Mean = 83.97 and 68.79; Standard Deviation = 15.65 and 14.46 for IMPROVE and control groups respectively; $F(1,62) = 11.46$, $p<.001$). Furthermore, analyses of videotaped interactions that recoded the actual teaching of an IMPROVE teacher and a control-group teacher indicated that the IMPROVE teacher was better able than her counterpart in the control group to guide her students in activating metacognitive processes, and to promote students' understanding.

Figure 8.2. **Impact of IMPROVE on teachers' mathematical knowledge**

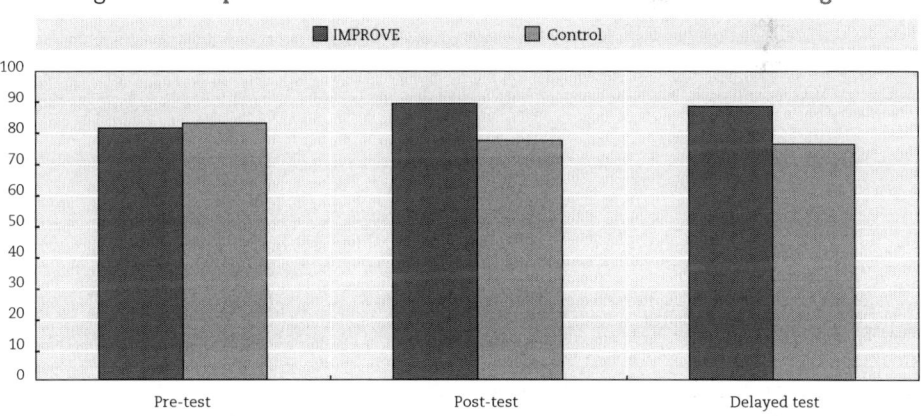

StatLink ᵐˢᵖ http://dx.doi.org/10.1787/888933149204

Source: Kramarski and Revach (2009), "The challenge of self-regulated learning in mathematics teachers' professional training", *Educational Studies in Mathematics*, Vol. 72(3), pp.379-399.

Figure 8.3. Impact of IMPROVE on teachers pedagogical content knowledge

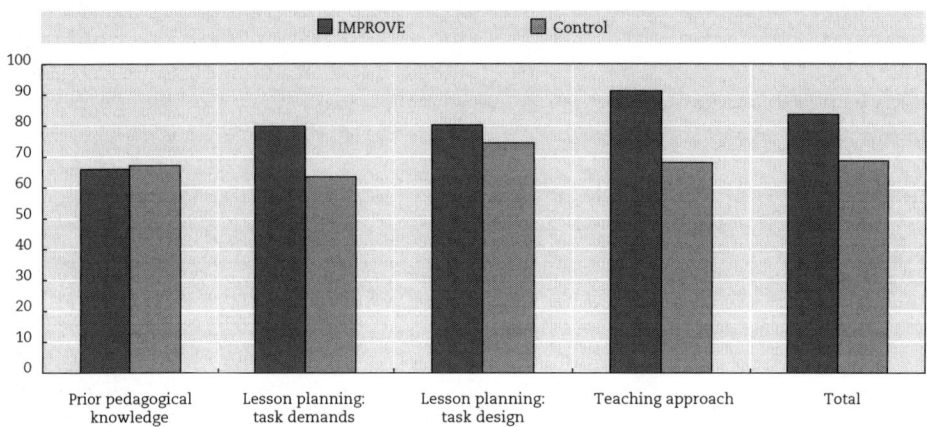

StatLink ⇲ http://dx.doi.org/10.1787/888933149214

Source: Kramarski and Revach (2009), "The challenge of self-regulated learning in mathematics teachers' professional training", *Educational Studies in Mathematics*, Vol. 72(3), pp.379-399.

These findings enlarge the findings of previous studies showing the effects of IMPROVE on schooling outcomes. While the studies reviewed in previous chapters showed the positive effects of IMPROVE on students' CUN and routine problem solving, this study shows similar effects on teachers' mathematics and pedagogical knowledge as well as on lesson planning. Interestingly, for both students and teachers, the effects of IMPROVE were evident not only on the immediate assessments, but also on a delayed high-stakes exam (see also Chapter 5, sections on high-stake situations and on immediate and lasting effects).

Teachers' judgment of learning (JOL)

If indeed professional development programmes require teachers to play the roles of learners and teachers simultaneously, it would be interesting to examine how teachers judge their learning in these settings, and whether or not they expect to apply what they have learned in the course in their work as teachers. Mevarech and Shabtay (2012) addressed these issues.

Judgment of learning (JOL) is a unique component of metacognition. It refers to individuals' capabilities to judge the extent to which they would recall an assigned task after a certain time, be it an hour, a day, a week, a month or a year. For example, learners are asked to predict the likelihood that they will recall the definition of a parallelogram next year, or whether they would be able to solve a speed-time-distance problem next month. Learners base their monitoring and control processes on their judgment of learning. When learners think that they are not prepared enough for an exam, they will allocate more study time and will probably also decide which strategies to apply in order to improve learning.

Research in psychology has placed a lot of emphasis on understanding JOL, its mechanism and its impact on memory. In a series of studies conducted in lab situations, researchers (e.g. Koriat, 2008) showed that when participants estimate the materials to be learned to be easy, their JOL is high and they think that they would easily recall these materials. Koriat called these findings: "easy to learn – easy to recall".

The way learners judge their learning applies to students in all ages, including teachers who participate in professional development programmes. If teachers in an in-service professional development programme think that they are unlikely to remember the material taught in the course, the probability that they will apply that material in their classroom is rather slim.

Given the high costs of professional development programmes, policy makers are often interested in how teachers judge their learning in such programmes. Hence, it is important to identify the conditions that would increase teachers' JOL.

Mevarech and Shabtay (2012) designed a study in which they examined teachers' JOL under different conditions. In this study, primary school mathematics teachers who participated in a professional development programme were randomly assigned into one of two groups taught by the same instructor: one group studied via IMPROVE, and the other in a traditional way with no explicit metacognitive intervention. Towards the end of the course, and again about a month later, the two groups were asked to judge their learning. The JOL questionnaire asked the participants to estimate the extent to which they are sure that they would remember a month from "now" the mathematical concepts, tasks and specific examples given during the course. An open-ended questionnaire assessed the accuracy of the teachers' judgment of learning by asking them to give examples of the concepts, tasks and exercises given in the course. The open-ended part followed the JOL questionnaire.

The findings indicate that while no significant differences were found between the groups prior to the beginning of the study, the IMPROVE group scored higher on JOL and were also more accurate in their judgment of learning than the control group. Even more noteworthy is the finding that while only 16% of the control group claimed that they would apply what they learned in the course in their classrooms, about 70% (68.6%) of the IMPROVE teachers declared that they would do so. Figure 8.4 presents the mean scores for the teachers' judgment of learning at the end of the course and after one month; Figure 8.5 presents the teachers' accuracy of judgment of learning with regard to the concepts, tasks and examples taught in the professional development course.

In summary, the studies reviewed above showed the positive effects of applying metacognitive scaffolding to professional development programmes on teachers' mathematics knowledge, JOL, and PCK. However, all these studies were implemented in traditional classrooms with no ICT. Would similar findings be found in professional development programmes emphasising the use of ICT? This issue is reviewed below.

Figure 8.4. **Impact of IMPROVE on judgement of learning**

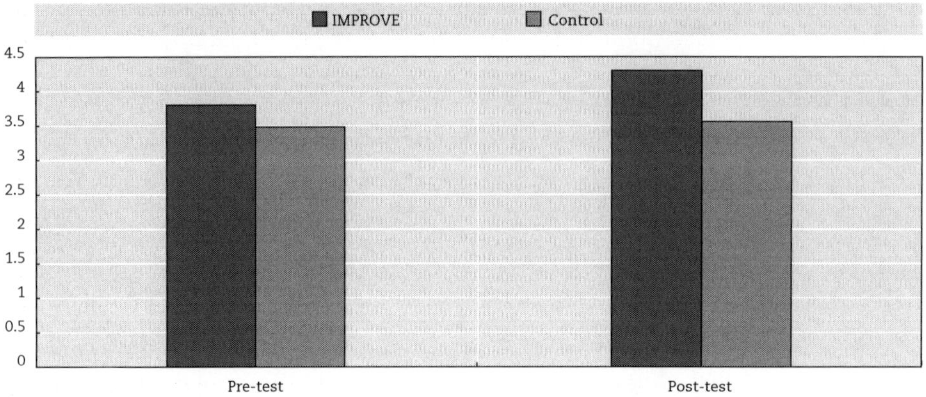

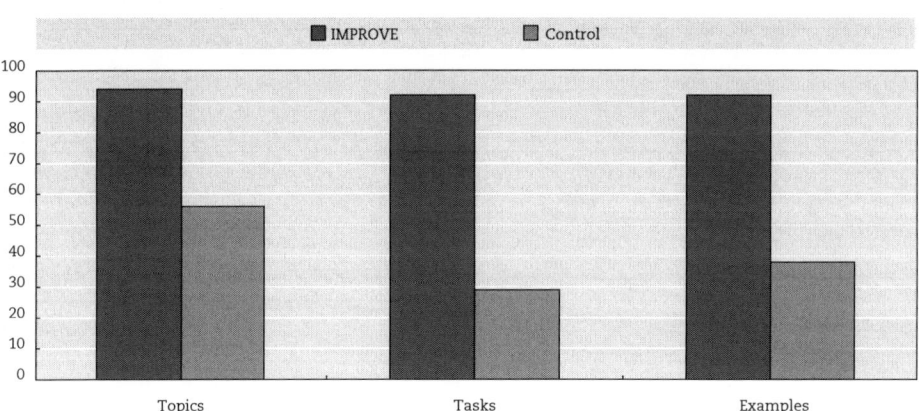

StatLink http://dx.doi.org/10.1787/888933149228

Source: Mevarech and Shabtay (2012), "Judgment-of-learning and confidence in mathematics problem solving: A metacognitive benefit for the explainer", in *Metacognition 2012 – Proceedings of the 5th Biennial Meeting of the EARLI Special Interest Group 16 Metacognition*, Milan, 5-8 September.

Figure 8.5. **Impact of IMPROVE on the accuracy of teachers' judgement of learning**

StatLink http://dx.doi.org/10.1787/888933149232

Source: Mevarech and Shabtay (2012), "Judgment-of-learning and confidence in mathematics problem solving: A metacognitive benefit for the explainer", in *Metacognition 2012 – Proceedings of the 5th Biennial Meeting of the EARLI Special Interest Group 16 Metacognition*, Milan, 5-8 September.

Teachers' professional growth in professional development programmes based on ICT

Given the strengths and limitations of ICT environments (see Chapter 7), Phelps, Graham and Kerr (2004) proposed embedding a metacognitive approach into these environments so that "rather than specific objectives or outcomes being 'imposed' on learners, participants are encouraged to identify, articulate and pursue personally relevant goals, including those related to skills, attitudes, confidence, values and understandings, integration and school leadership" (Phelps, Graham and Kerr, 2004, p. 49).

A sample of 40 secondary-school teachers, representing 7% of the secondary school teachers in a district, participated in a study that took place during a professional development course that extended over a period of two school terms. Participants attended two workshops that included interaction with self-paced print, compact disk and web-based resources, and participated in online communication. The metacognitive process was scaffolding though the first workshop and a print-based "Thinking module". Teachers were tested before and after the intervention.

Phelps, Graham and Kerr (2004) reported that embedding the ICT within the metacognitive pedagogy significantly improved participants' computer skills development, and influenced the teachers' willingness to apply their metacognitive learning and reflection not only to their own professional development but also to their interactions with their students and fellow teachers in their school (p. 56). Furthermore, the majority of the teachers (37 out of 40 teachers) were positive or highly positive about the use of metacognitive processes. Similar findings were also reported by Bayer (2002) and Vrieling, Bastiaens and Stijnen (2012) who also investigated the effects of metacognitive pedagogy embedded in professional development programme on the use of ICT. Phelps et al. concluded:

These findings indicate that the metacognitive approach has broader outcomes and implications than as simply as approach to ICT professional development. Rather, ICT is used as a medium to engage teachers in confronting broader issues about their own, their students', and their fellow teachers' learning, The metacognitive approach also actively fostered the formation of support structures and networks which could support teachers' learning beyond their involvement in the professional development initiative. As such it becomes a powerful vehicle to support change processes within the school environment (Phelps, Graham and Kerr, 2004).

The effects of metacognitive pedagogies on pre-service teachers

Parallel to the implementation of metacognitive pedagogies in in-service professional development programmes, various attempts have also been made to apply this approach to pre-service teacher training. These studies were also implemented in both ICT or non-ICT environments. While the ICT-based professional development programmes mainly aimed to improve the general use of technology, the non-ICT programmes focused on the teaching of specific domains, such as maths or science. The use of different kinds of ICT (e.g. hypermedia, e-learning, web-based) led researchers to design studies that assessed its effects on various outcomes, including PCK, SRL, SE, or use of ICT.

Studies into the effects of including metacognitive scaffolding into ICT-based professional development programmes have followed two lines of research. The first examines the effects of professional development courses based on the use of a specific technology. The second compares e-learning programmes with traditional ones in face-to-face settings.

Adding metacognitive scaffolding to a single-technology course

Kramarski and Michalsky (2010) conducted a study with 95 pre-service secondary school science teachers who were enrolled in a mandatory first-year course "Designing Learning Activities with a Hypermedia Environment". Participants were randomly assigned into one of two conditions: using hypermedia either with or without metacognitive scaffolding (N = 47 and 48, respectively). The outcomes were evaluated along two dimensions: technology pedagogical content knowledge (TPCK) and self-regulated learning. TPCK referred to the comprehension of a structured study unit and the capability to use the technology to design a two-lesson study unit on the topic of "the effects of drugs on people's lives". SRL was assessed by the 50-item Motivated Strategies for Learning Questionnaire (MSLQ; Pintrich, Smith, Garcia and McKeachie, 1991). The MSLQ measured the three SRL components: cognition, metacognition and motivation, all adapted to the pedagogical context.

The results indicated that the Hypermedia + Metacognition group significantly outperformed the Hypermedia only students on both TPCK (comprehension and lesson design) and SRL (cognition, metacognition, and motivation). Figures 8.6 and 8.7 present the mean scores on TPCK and SRL by time and conditions.

Figure 8.6. **Effect of metacognitive scaffolding on technology pedagogical content knowledge**

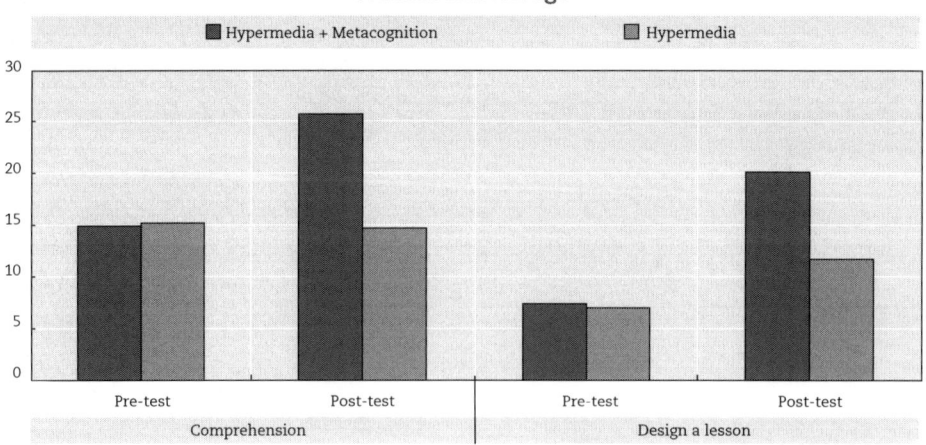

StatLink ⬛🔗 http://dx.doi.org/10.1787/888933149246

Source: Kramarski and Michalsky (2010), "Preparing preservice teachers for self-regulated learning in the context of technological pedagogical content knowledge", Learning and Instruction, Vol. 20(5), pp. 434-447.

Figure 8.7. **Effect of metacognitive scaffolding on self-regulated learning**

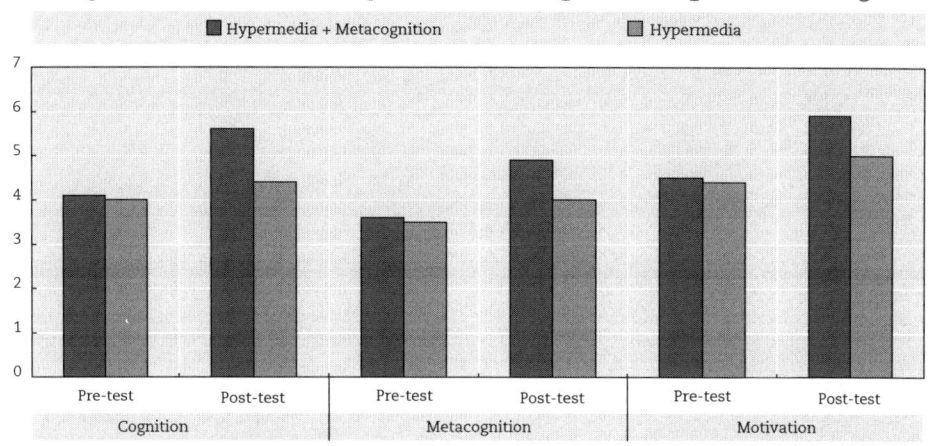

StatLink 🔗 http://dx.doi.org/10.1787/888933149255

Source: Kramarski and Michalsky (2010), "Preparing preservice teachers for self-regulated learning in the context of technological pedagogical content knowledge", *Learning and Instruction*, Vol. 20(5), pp. 434-447.

While Kramarski and Michalsky looked at the ICT environment as a whole, Kauffman, Ge, Xie and Chen (2008) investigated how the manipulation of problem-solving and self-reflection prompts affected pre-service teachers' ability to solve complex simulated pedagogical cases in a web-based learning environment. They implemented a 2x2 study design in which one factor was problem-solving prompts (applied or not applied) and the other reflection prompts (applied or not applied). The problem-solving questions included: "What do you see as the primary problem? Why is that a problem? And what do you see as possible solutions to this problem?" The reflection question was "How certain are you that you have identified the primary problem?" Measures included participants' analyses of two case studies presenting authentic scenarios that describe a teacher having difficulties related to classroom management. The participants' written responses were assessed for clarity, flexibility, argument development and a total evaluation.

The findings indicated that pre-service teachers who received both problem-solving and reflection prompts outperformed the other groups on clarity, fluency, argumentation, and total evaluation, but no significant differences were found between the other three groups. Figure 8.8 presents the mean scores by condition.

It seems, therefore, that "prompting students with questions designed to support their use of problem solving strategies (and perhaps other strategies) is an effective technique for helping students successfully navigate through traditional Web-based environments" (Kauffman, Ge, Xie and Chen, 2008, p.132). The prompts provide guidance in identifying the goals, monitoring the progress, and reflecting on the outcomes. Kauffman et al. concluded:

> Clearly, the problem solving prompts had a positive influence on students writing and thinking about the problem, but only if the problem solving

prompts were followed by reflection prompts... (In parallel) providing students with an opportunity to reflect on their own work is an effective technique for improving problem solving and achievement, but only when accompanied by a clear understanding of the problem solving process. Thus, teachers and instructional designers should make every opportunity to not only clearly identify the learning task, but also to help students know when and how to reflect on their work before submitting their solution (Kauffman, Ge, Xie and Chen, 2008, p. 133).

Figure 8.8. **Impact of problem-solving and reflection prompts on case analysis**

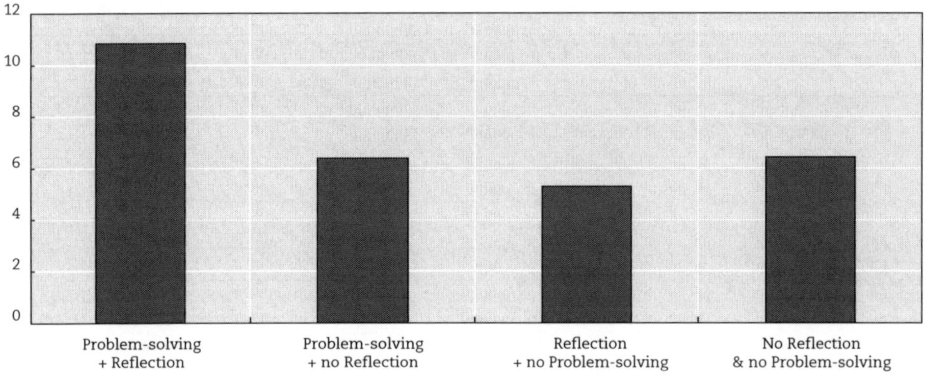

StatLink http://dx.doi.org/10.1787/888933149264

Source: Source: Kauffman, Ge, Xie and Chen (2008), "Prompting in web-based environments: Supporting self-monitoring and problem solving skills in college students", *Journal of Educational Computing Research*, Vol. 38(2), pp. 115-137.

Comparing e-learning with face-to-face settings

The second line of research compared the effects of professional development programmes based on e-learning (EL) with traditional human learning in face-to-face (F2F) settings with and without metacognitive scaffolding. This creates four learning conditions: EL with metacognition, F2F with metacognition, EL with no metacognition and F2F with no metacognition. Kramarski and Michasky (2009) compared the effects of these four conditions on 194 pre-service teachers. Their professional growth was assessed by MSLQ (Pintrich et al., 1991), adapted to the pedagogical context (cognition, metacognition, and motivation), and PCK (designing a learning unit). The SRL scores for each component ranged from 1 to 7; PCK scores ranged from 1 to 24.

The results indicated that students who were exposed to the metacognitive scaffolding either in an e-learning or face-to-face environment outperformed their counterparts who studied in the correspondence environments with no metacognitive scaffolding. Figure 8.9 presents the mean scores on self-regulated learning (cognition, metacognition, and motivation) by time and condition, and Figure 8.10 presents PCK total mean scores by time and conditions.

Figure 8.9. **Effect of metacognition on self-regulated learning in e-learning and face-to-face settings**

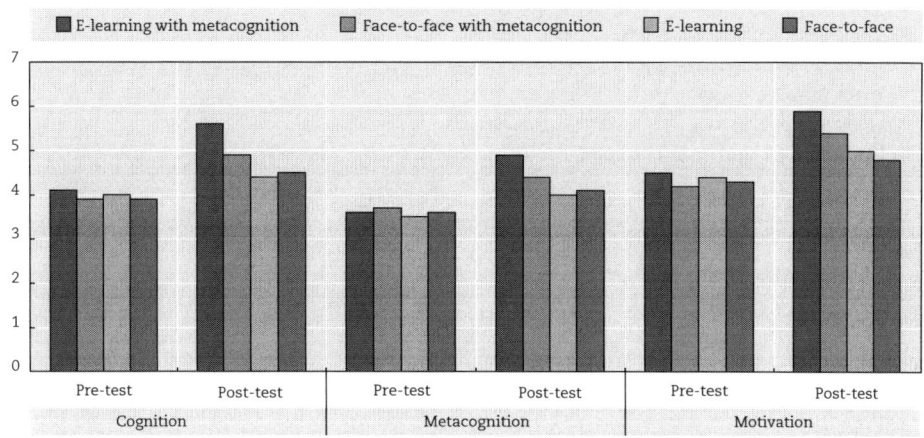

StatLink http://dx.doi.org/10.1787/888933149277

Source: Kramarski, B and T. Michalsky (2010), "Preparing preservice teachers for self-regulated learning in the context of technological pedagogical content knowledge", *Learning and Instruction,* Vol. 20(5), pp. 434-447.

Figure 8.10. **Effect of metacognition on pedagogical content knowledge in e-learning and face-to-face settings**

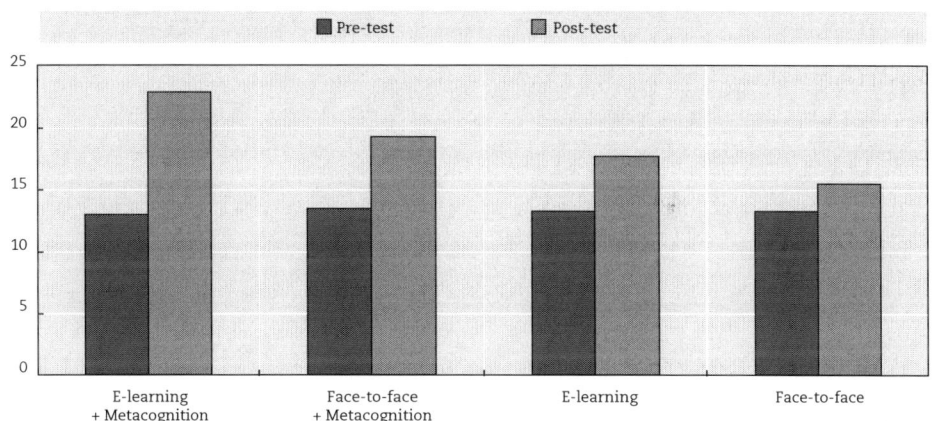

StatLink http://dx.doi.org/10.1787/888933149284

Source: Kramarski, B and T. Michalsky (2010), "Preparing preservice teachers for self-regulated learning in the context of technological pedagogical content knowledge", *Learning and Instruction,* Vol. 20(5), pp. 434-447.

These findings are in line with the findings of Zion, Michasky and Mevarech (2005) who examined the differential effects of the same four learning environments on high school students' science literacy (see Chapter 7, Section on asynchronous learning networks). In both studies, the exposure to the metacognitive scaffolding in either e-learning or face-to-face environments produced the highest mean scores for PCK and science literacy, respectively.

Using metacognition in traditional settings

Researchers have also examined the effects of professional development programmes implemented in traditional settings with no ICT. In particular, researchers have studied how the manipulation of different kinds of metacognitive prompts could make a difference to prospective teachers' professional growth. A series of studies looking into the "black box" of pre-service professional programmes took this approach. For example, in response to the hypothesis that effective self-regulation support should promote students' *skills* (metacognition) and *will* (motivation), Kohen and Kramarski (2012) investigated the effects of self-regulation support combining metacognition and motivation components on student teachers' motivation, metacognition and pedagogical content knowledge.

A total of 97 participants were engaged in a mandatory "micro-teaching" course based on role-play simulations in which the pre-service teachers played the role of a teacher in front of their peers, who played the role of students. Each participant's "teaching activity" lasted 10 minutes followed by feedback given by the peer students and the instructor.

Participants were randomly assigned into one of two groups: *reflective support* (RS) in which self-regulation scaffolding was provided, or *no support* (a control group). The reflective support group was explicitly exposed to the use of self-regulation applied by IMPROVE (Mevarech and Kramarski, 1997) and motivational aspects (Efklides, 2008; 2011; Pintrich, 2000) and practised pedagogical exercises. Self-regulation was discussed by focusing on *what* aspects are important and *why*, and *how* and *when* to implement them in classrooms. The metacognitive aspect referred to increasing students' motivation, through means such as using exciting lesson openings or interesting demonstrations. The group with no reflective support was exposed to theoretical pedagogical frameworks (Shulman, 1986) and discussed pedagogical issues (e.g. the structure of the lesson and student-teacher interaction).

The student teachers' teaching experiences were videotaped and transcribed. The self-regulation process was assessed by using a coding scheme to code events during real-time teaching behaviour according to two major aspects: 1) metacognition (planning, information management, monitoring, debugging and evaluation); and 2) affective variables (motivation, self-efficacy and teaching anxiety). The motivation aspect was assessed by the Motivated Strategies for Learning Questionnaire (MSLQ; Pintrich et al., 1992).

The results indicated that the group exposed to reflective support displayed higher mean scores on metacognitive, self-efficacy and motivational behaviour. In addition, the Reflective Support group manifested less teaching anxiety than the group with no reflective support (Figures 8.11 and 8.12). The findings are in line with previous studies showing the effects of the combined approach on cognitive and affective outcomes (see Chapter 6).

Figure 8.11. **Effect of Reflective Support on planning, processing, monitoring and debugging**

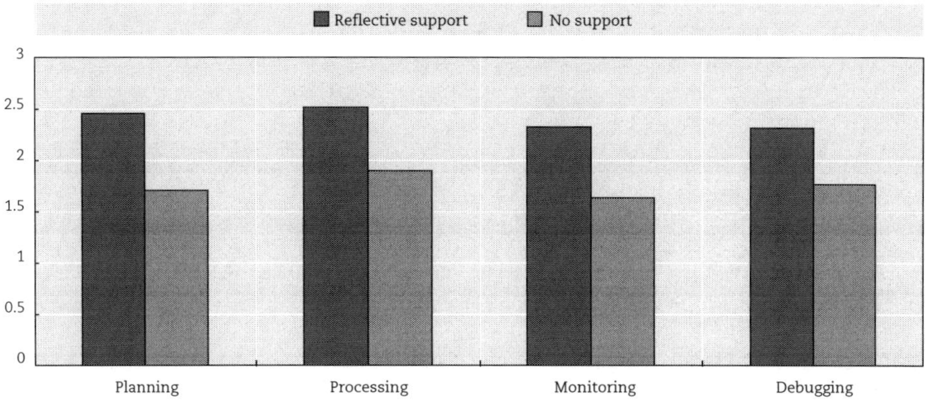

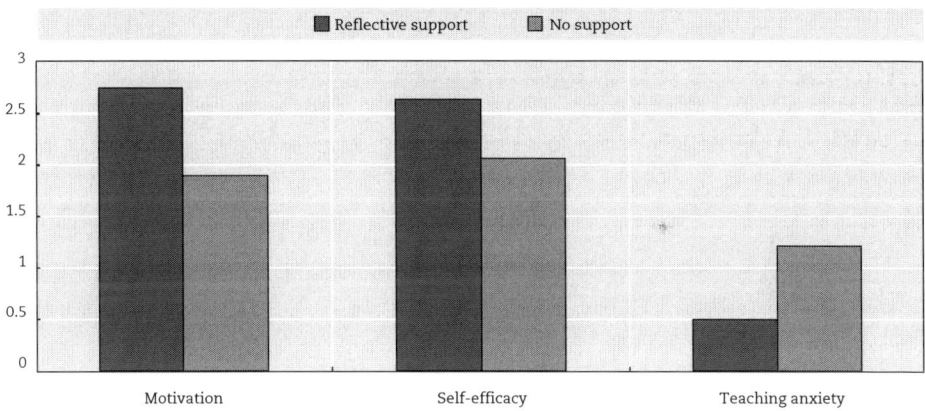

StatLink ⟐ http://dx.doi.org/10.1787/888933149297

Source: Kohen and Kramarski (2012), "Developing self-regulation by using reflective support in a video-digital microteaching environment", *Education Research International*, Vol. 2012, Article ID 105246, *http://dx.doi.org/10.1155/2012/105246*

Figure 8.12. **Effect of Reflective Support on motivation, self-efficacy and teaching anxiety**

StatLink ⟐ http://dx.doi.org/10.1787/888933149304

Note: A higher score on teacher anxiety indicates higher anxiety.

Source: Kohen and Kramarski (2012), "Developing self-regulation by using reflective support in a video-digital microteaching environment", *Education Research International*, Vol. 2012, Article ID 105246, *http://dx.doi.org/10.1155/2012/105246*.

Although this study showed promising findings for the effects of reflective support, the design of the study is based on the combination of two metacognitive prompts (cognition and motivation). This design eliminated the possibility of examining the specific contribution of each component by itself on pre-service teachers' professional growth. Michalsky (2012) hypothesised that there might be differences between pre-service teachers who are exposed to metacognitive scaffolding that focused on the cognitive component (Cog), those exposed to metacognitive scaffolding that enhanced motivation (Mot), and those exposed to metacognitive scaffolding that emphasises both cognitive and motivation components (CogMot). She compared these three groups to a fourth group with no metacognitive scaffolding (No Meta). The metacognitive scaffolding was based on a version of IMPROVE that was modified to fit the purposes of each condition. Table 8.2 presents these modifications.

Table 8.2. **IMPROVE self-questioning types and their SRL components embedded in PCK tasks**

Addressed Question	SRL Component	
	Cognitive-Metacognitive	Motivational
Comprehension	What is the phenomenon all about? What is the problem/task needing investigation?	What is your motivation for solving the problem/task? Explain. What will you do if you run into difficulties?
Connection	What do you already know about the phenomenon? What are the similarities/differences between the problem/task at hand and the problems/tasks you have encountered in the past? Please explain your reasoning.	What are the similarities/differences between your motivation/efforts/self-efficacy in the problem/task at hand and in the problems/tasks you have solved in the past? WHY?
Strategy	What are the inquiry strategies that are appropriate for solving the problem/task?	When/how should you implement a particular motivation strategy to solve the problem/task? What motivational strategies are appropriate for solving the problem/task?
Reflection	Does the solution make sense? Can you design the experiment/task in another way? How? Please explain your reasoning?	Do you feel good about your motivation/efforts/self-efficacy while comprehending the problem/task? Can you motivate yourself in another way? How? Explain.

The participants in this study were 188 pre-service science teachers who were randomly assigned into one of the above four conditions. All participants were pre- and post-tested on their professional growth assessed in this study along three dimensions: self-regulated learning (SRL) in a science pedagogical context, pedagogical metacognitive knowledge, and self-efficacy in teaching science.

The results indicated that all three metacognitive groups outperformed the control group with no metacognitive scaffolding on all measures of professional growth. Furthermore, within the metacognitive groups, the combined group (CogMet) attained the highest mean scores on all measures, whereas no significant differences were found between the metacognitive scaffolding based on one component (cognitive or motivational). Figures 8.13, 8.14 and 8.15 present the mean scores for SRL (cognition, metacognition, and motivation), metacognitive knowledge (declarative, procedural and conditional), and self-efficacy by time and condition.

Figure 8.13. **Effect of different interventions on self-regulated learning among pre-service teachers**

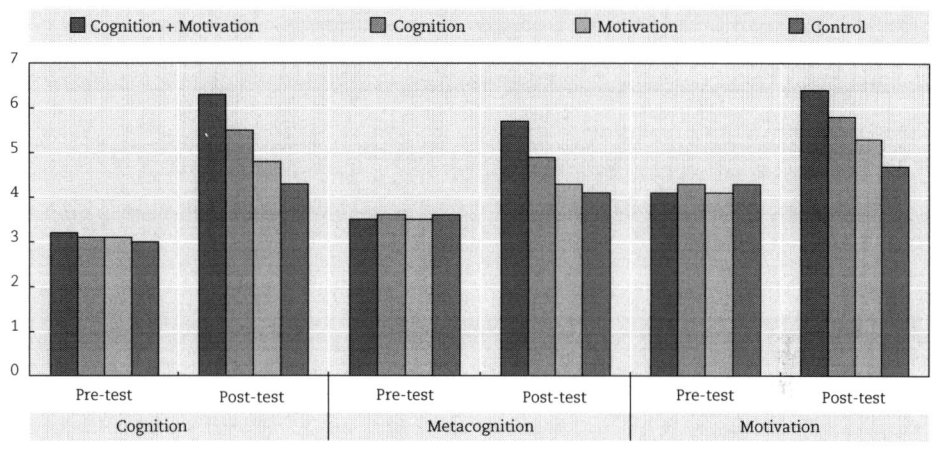

StatLink ᴍᵍᶰ http://dx.doi.org/10.1787/888933149313

Source: Michalsky (2012), "Shaping self-regulation in science teachers' professional growth: Inquiry skills", *Science Education*, Vol. 96(6), pp. 1106-1133.

Figure 8.14. **Effect of different interventions on metacognitive knowledge among pre-service teachers**

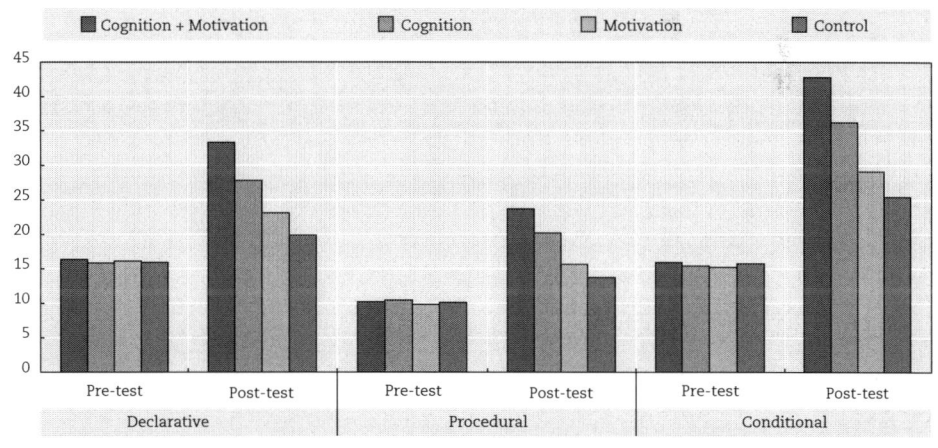

StatLink ᴍᵍᶰ http://dx.doi.org/10.1787/888933149320

Source: Michalsky (2012), "Shaping self-regulation in science teachers' professional growth: Inquiry skills", *Science Education*, Vol. 96(6), pp. 1106-1133.

Figure 8.15. **Effect of different interventions on self-efficacy among pre-service teachers**

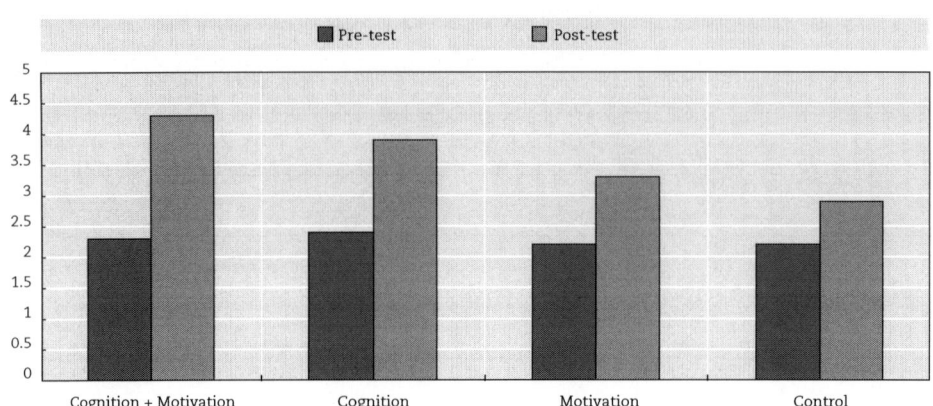

StatLink http://dx.doi.org/10.1787/888933149337

Source: Michalsky (2012), "Shaping self-regulation in science teachers' professional growth: Inquiry skills", *Science Education*, Vol. 96(6), pp. 1106-1133.

These findings are in line with Michalsky's previous study with 10th grade students (see Chapter 6, section on self-efficacy). In both studies, the provision of metacognitive pedagogy based on either cognition (problem solving) or motivation was found to be a necessary condition but not sufficient for improved achievement, whereas the combination of the two components had much stronger effects than each component by itself. The fact that the outcomes of these two studies are similar increases the likelihood that these findings can be generalised to other settings that also employ a modified version of metacognitive scaffolding (e.g. IMPROVE) in which the cognitive and motivational metacognitive components are combined.

Conclusion

The studies reviewed in this chapter showed how metacognitive scaffolding plays an essential role in professional development programmes for both pre- and in-service teachers. Metacognitive interventions enhance the effects of these programmes on teachers' professional growth as indicated by assessments of their PCK, metacognition, judgment of learning, confidence and self-efficacy. Hence, these findings have important implications for the design of these programmes:

- Asking teachers to reflect on their work has positive effects on their own and their students' problem-solving skills. Hence, professional development programmes, whether or not they are based on ICT, might shift from focusing only on the content and skills to be taught towards restructuring the environments to apply metacognitive pedagogies that enrich teachers' professional growth.

- Metacognitive pedagogies that combine both cognitive and motivational components seem to be more effective than those applying only one component.

Because the current existing metacognitive pedagogies such as IMPROVE are usually based on generic self-addressed questions, it should be easy to modify them to include both components.

- As "learning by doing" and "learning from experience" are effective training methods, professional development programmes based on the provision of metacognitive scaffolding should be restructured accordingly.

- None of the studies reviewed in this chapter followed the teachers into their classrooms to discover the extent to which teachers indeed implement what they have learned in the professional development programmes studied. This issue will be essential for evaluating the effectiveness of pre- and in-service professional development programmes, and it definitely merits future research.

As the field of metacognition in general and metacognitive pedagogies in particular continues to develop, professional development programmes should be informed by this research and be updated accordingly.

References

Artzt, A.F., and E. Armour-Thomas (1998), "Mathematics teaching as problem solving: A framework for studying teacher metacognition underlying instructional practice in mathematics", *Instructional Science*, Vol. 26, pp. 5-25.

Efklides, A. (2011), "Interactions of metacognition with motivation and affect in self-regulated learning: The MASRL model", *Educational Psychologist*, Vol. 46(1), pp. 6-25.

Efklides, A. (2008), "Metacognition: Defining its facets and levels of functioning in relation to self-regulation and co-regulation", *European Psychologist*, Vol. 13(4), pp. 277-287.

Hurme, T.-R., K. Merenluoto and S. Järvelä (2009), "Socially shared metacognition of pre-service primary teachers in a computer-supported mathematics course and their feelings of task difficulty: A case study", *Educational Research and Evaluation*, Vol. 15(5), pp. 503-524.

Kauffman, D.F., X. Ge., K. Xie and C.H. Chen (2008), "Prompting in web-based environments: Supporting self-monitoring and problem solving skills in college students", *Journal of Educational Computing Research*, Vol. 38(2), pp. 115-137.

Kohen, Z. and B. Kramarski (2012), "Developing self-regulation by using reflective support in a video-digital microteaching environment", *Education Research International*, 10 pages, *http://dx.doi.org/10.1155/2012/105246*.

Kolb, D.A. (1984), *Experiential Learning: Experience as the Source of Learning and Development*, Prentice-Hall, New Jersey.

Koren, M. (2008), "The effects of metacognitive guidance on teachers' mathematics content knowledge and pedagogical knowledge: The case of polygons' areas and perimeters", MA Thesis, Bar-Ilan University (Hebrew).

Koriat, A. (2008), "Easy comes, easy goes? The link between learning and remembering and its exploitation in metacognition", *Memory and Cognition*, Vol. 36(2), pp. 416-428.

Kramarski, B. (2008), "Promoting teachers' algebraic reasoning and self-regulation with metacognitive guidance", *Metacognition and Learning*, Vol. 3(2), pp. 83-99.

Kramarski, B and T. Michalsky (2010), "Preparing preservice teachers for self-regulated learning in the context of technological pedagogical content knowledge", *Learning and Instruction*, Vol. 20(5), pp. 434-447.

Kramarski, B. and T. Michalsky (2009), "Investigating pre-service teachers' professional growth in self-regulated learning environments", *Journal of Educational Psychology*, Vol. 101(1), pp. 161-175.

Kramarski, B. and T. Revach (2009), "The challenge of self-regulated learning in mathematics teachers' professional training", *Educational Studies in Mathematics*, Vol. 72(3), pp. 379-399.

Mevarech, Z.R. and B. Kramarski (1997), "IMPROVE: A ,ultidimensional method for teaching mathematics in heterogeneous classrooms", *American Educational Research Journal*, Vol. 34(2), pp. 365-395.

Mevarech, Z.R. and G. Shabtay (2012), "Judgment-of-learning and confidence in mathematics problem solving: A metacognitive benefit for the explainer", in *Metacognition 2012 – Proceedings of the 5th Biennial Meeting of the EARLI Special Interest Group 16 Metacognition*, Milan, 5-8 September.

Michalsky, T. (2013), "Integrating skills and wills instruction in self-regulated science text reading for secondary students", *International Journal of Science Education*, Vol 35(11), pp. 1846-1873.

Michalsky, T. (2012), "Shaping self-regulation in science teachers' professional growth: Inquiry skills", *Science Education*, Vol. 96(6), pp. 1106-1133.

Nool, N.R. (2012), "Exploring the metacognitive processes of prospective mathematics teachers during problem solving", in *2012 International Conference on Education and Management Innovation*, IPEDR Vol. 30, IACSIT Press, Singapore, pp. 302-306.

Perry, N.E., L. Phillips and L. Hutchinson (2006), "Mentoring student teachers to support self-regulated learning", *The Elementary School Journal*, Vol. 106(3), pp. 237-254.

Phelps, R., A. Graham and B. Kerr (2004), "Teachers and ICT: Exploring a metacognitive approach to professional development", *Australasian Journal of Educational Technology*, Vol. 20(1), pp. 49-68.

Pintrich, P.R. (2000), "The role of goal orientation in self-regulated learning", in M. Boekaerts, P.R. Pintrich and M. Zeidner (eds.), *Handbook of Self-Regulation*, Academic Press, San Diego, pp. 451-502.

Pintrich, P.R., D.A.F. Smith, T. Garcia and W.J. McKeachie (1991), *A Manual for the Use of the Motivational Strategies Learning Questionnaire (MSLQ)*, National Center for Research to Improve Postsecondary Teaching and Learning, University of Michigan, Ann Arbor, MI.

Prytula, M.P. (2012), "Teacher metacognition within the professional learning community", *International Education Studies*, Vol 5(4), *http://dx.doi.org/10.5539/ies.v5n4p112*.

Putnam, R.T. and H. Borko (2000), "What do new views of knowledge and thinking have to say about research on teacher learning?" *Educational Researcher*, Vol. 29(1), pp. 4-15.

Randi, J. and L. Corno (2000), "Teacher innovations in self-regulated learning", in M. Boekaerts, P.R. Pinrich and M. Zeidner (eds.), *Handbook of Self-Regulation*, Academic Press, San Diego, pp. 651-685.

Shulman, L.S. (1986), "Those who understand: Knowledge growth in teaching", *Educational Researcher,* Vol. 15(2), pp. 4-14.

Veenman, M.V.J., B.H.A.M. Van Hout-Wolters and P. Afflerbach (2006), "Metacognition and learning: Conceptual and methodological considerations", *Metacognition and Learning,* Vol. 1(1), pp. 3-14.

Verschaffel, L., F. Depaepe and E. De Corte (2007), "Upper elementary school teachers' conceptions about and approaches towards mathematical modelling and problem solving: How do they cope with reality?", paper presented at the Conference on Professional Development of Mathematics Teachers Research and Practice from an International Perspective, held at the Mathematische Forschungsinstitut Oberwolfach, Germany.

Wilburne, J.M. (1997), "The effect of teaching metacognition strategies to pre-service elementary school teachers on their mathematical problem-solving achievement and attitude", unpublished PhD Thesis, Temple University.

Wilson, N.S. and H. Bai (2010), "The relationships and impact of teachers' metacognitive knowledge and pedagogical understandings of metacognition", *Metacognition and Learning,* Vol. 5(3), pp. 269-288.

Zion, M., T. Michalsky and Z.R. Mevarech (2005), "The effects of metacognitive instruction embedded within an asynchronous learning network on scientific inquiry skills", *International Journal of Science Education,* Vol. 27(8), pp. 957-983.

Zohar, A. (1999), "Teachers' metacognitive knowledge and the instruction of higher order thinking", *Teaching and Teacher Education,* Vol. 15(4), pp. 413-429.

Chapter 9

Looking backwards:
Summary and conclusion

This chapter summarises the main findings of the book and concludes.

In the last decade it has become almost a truism that the major goal of mathematics education is to develop quantitatively literate citizens who have the capability to "analyse, reason and communicate effectively as they pose, solve, and interpret mathematical problems in a variety of situations involving quantitative, spatial, probabilistic or other mathematics concepts" (OECD, 2004, p. 34). The OECD further indicates that the goal of education in innovation-driven societies is not to train students to become professional mathematicians, but rather to develop students' abilities to possess mathematics knowledge and understanding, apply the knowledge and skills in solving problems, and be able to make decisions based on quantitative information. Innovating to learn and learning to innovate has become a crucial issue in critical maths for the 21st century (OECD, 2008a).

In addressing these issues, many researchers, educators and policy makers are concerned with two basic questions. First, what are the maths problems and sets of skills that are useful for developing literate citizens in innovation-driven societies? Second, and more challenging, which methods are effective for promoting these sets of skills, and is there evidence that certain pedagogies actually work in our highly structured compulsory educational systems?

For the first question, there is a broad consensus that in addition to learning to solve routine mathematics problems, students have to become acquainted with the skills and processes that are appropriate for solving complex, unfamiliar and non-routine (CUN) tasks as well as authentic problems. There is also a great degree of agreement that the inclusion of metacognition is necessary for solving CUN tasks. Metacognition – thinking about and regulating thinking – is a powerful tool for enhancing learning.

As to the second question, it is not at all self-evident how best to promote the skills needed to solve CUN tasks. The debates are so heated that some commentators have referred to them as "maths wars". On the one hand, some advocate the return to basic skills and traditional teaching methods. On the other, educators proclaim the need to put into practice the advantages of progressive, inquiry-based pedagogies, particularly for enhancing the solution of CUN tasks. For years, the pedagogical pendulum has swung from one side of the debate to the other. The inclusion of metacognition might leave the pendulum in the middle by providing evidence-based instructional methods that connect the different approaches emphasizing the importance of "algorithmic fluency" side-by-side with conceptual understanding, reasoning, problem solving, creativity and communication in mathematics. Practicing these processes via metacognition deepens conceptual understanding and contributes to the development of solid base mathematical knowledge that most students cannot attain by merely drilling the algorithms. The question, thus, is not "either-or"; it is not either algorithm mastery or conceptual understanding; it is also not the issue of drilling vs. "discovering" or "inventing" the standard algorithms. Both algorithm fluency and conceptual understanding of maths could be attained by teaching students to use metacognition: how to comprehend problems, construct bridges for applying the knowledge and skills, use appropriate strategies, and reflect on all the stages of the solution.

Epistemologically, metacognition is the "engine" that enables the cognitive processes to function. During problem solving, information flows from the object-basis to the meta-level via monitoring and control processes. The "metacognitive engine" includes the "starter" for initiating and planning the cognitive process, the monitoring and control components that are activated during the problem solving, and the reflection that acts at all stages of the process but in particular at the end, when looking backwards is most relevant. The application of metacognitive processes is particularly beneficial in the solution of CUN tasks for which there are no ready-made algorithms, and therefore no automatic application is possible. Although people can sometimes solve maths problems by guessing or trial and error, these strategies are usually ineffective and often result in failure and frustration. Cognitive and metacognitive skills go hand in hand and thus need to be deliberately taught and practised together.

Surprisingly, although there is a wealth of research showing the positive effects of metacognition on mathematics problem solving even when ability has been controlled for, observations have indicated that teachers rarely use explicit metacognitive scaffolding in the classroom. The reasons for that might be twofold: 1) teachers simply do not know how to do so; and 2) most textbooks and teacher guides do not include metacognitive prompts.

The good news is that metacognition is teachable at all educational levels, from kindergarten to tertiary education. On the basis of these findings, researchers have started to design instructional methods that aim to promote metacognitive processes in mathematics problem solving. The five metacognitive pedagogical models reviewed in this book are those of Polya, Schoenfeld, IMPROVE, Verschaffel et al. and the Singapore model. All these models provide techniques for training students to use some form or another of self-directed metacognitive questioning in maths problem solving. These models work best in a co-operative learning environment (e.g. Slavin, 2010) where students study in small groups, articulate their mathematical reasoning, and describe their heuristics. In all of them too, the teacher plays an important role in explicitly modelling the use of metacognition.

The self-directed metacognitive questioning scaffolds the problem solving processes by prompting comprehension (e.g. what is the problem all about?), connection (e.g. how is the problem in hand similar to or different from what you have solved in the past), strategic (e.g. what strategies are appropriate for solving the problem?) and reflection (e.g. am I stuck, why? Can I solve the problem differently, how?). These self-directed metacognitive questions are generic in nature and thus could be easily modified to be used in other domains, such as science, reading, or even for fostering social-emotional outcomes (see below). For example, in learning science, the "problem" can be replace by "phenomenon", "maths strategies" by "inquiry techniques", etc.

Metacognitive pedagogies have been largely examined in the educational research arena. Among these methods, IMPROVE (Mevarech and Kramarski, 1997) is the one which has been most widely researched. Its effects on mathematics achievement have been studied on students from kindergarten to tertiary education, in both ICT-based and traditional learning environments, in co-operative and individualised settings, and in pre- and in-service professional teacher development programmes. From its very beginning, IMPROVE has proved to be an ecologically valid method, successfully implemented in "real" classrooms with ordinary teachers. Experimental and quasi-experimental studies indicate that IMPROVE students outperform their peers who studied with no metacognitive guidance on routine, authentic and various kinds of CUN tasks. In addition, in most studies, IMPROVE students were better able than their counterparts to articulate their mathematics reasoning. These positive effects were observed at all levels, among kindergarten children, primary, secondary and college students, and in high-stakes educational settings such as matriculation exams. The advantages were evident when the method was implemented over a short term, as well as a full academic year. Interestingly, the lasting effects of IMPROVE were still evident a year later even when IMPROVE students were no longer being exposed to metacognitive guidance. Moreover, a series of studies indicated the positive effects of IMPROVE when implemented in interactive learning environments (e.g. co-operative learning) with or without ICT. However, all these studies were carried out by the designers of the method and their teams, and some of these studies involved a relatively small sample of participants. It would be interesting to examine the effects of IMPROVE in other educational settings.

Studies by Schoenfeld (1987), Lester et al. (1989), Verschaffel et al. (1999) and many others also reported positive effects of metacognitive instruction on various types of CUN tasks. Overall, these studies indicate that metacognitive instructional methods have the potential to promote metacognition and the solution of CUN tasks of primary, secondary and tertiary students. Meta-analyses summarising dozens of studies involving thousands of students suggests that metacognitive interventions have strong positive effects on self-regulation and mathematics achievement (e.g. Dignath and Buettner, 2008; Dignath et al., 2008).

The findings further show the effects of IMPROVE and other similar metacognitive pedagogies on a variety of outcomes and populations. In general, the benefit that lower achievers derive from the metacognitive methodologies does not come at the expense of higher achievers. Additionally, the evidence indicates using the full set of self-directed metacognitive questions (comprehension, connection, strategic and reflection) provided by IMPROVE offers more benefit than using the strategic question by itself. The added value of implementing metacognitive instruction in parallel in mathematics and English classrooms was greater than just in mathematics alone. Lastly, IMPROVE has been successfully modified to also be used in science education.

Based on recent studies showing that cognition and emotions are inextricably linked in the brain (Hinton and Fischer, 2010), researchers have started to examine the possibilities of implementing metacognitive pedagogies for enhancing social-emotional outcomes in addition to cognitive achievements. In some of these studies, the social-emotional effects were considered as by-product of the originally designed interventions. In others, the self-directed metacognitive questioning was modified to facilitate the social-emotional outcomes. Still others aimed at enhancing both cognitive-metacognitive and emotional outcomes, and modified the metacognitive self-directed questioning accordingly. The findings indicated that IMPROVE and other similar methods enhanced students' motivation and self-esteem, as well as reducing maths anxiety.

The use of digital learning resources has the potential to go beyond textbooks (OECD, 2008b). Yet, quite often learners get lost in ICT environments because they are flooded with information that overloads their cognitive system. Metacognition and learning technologies have become a hot issue in education (Azevedo and Aleven, 2013). Reviewing studies in which metacognitive interventions supported various learning technologies (e.g. domain-specific software, asynchronous learning networks, cognitive tools, e-books, mobile learning through SMS, or hypermedia) indicates the beneficial effects of combining ICT and metacognition.

A modified version of IMPROVE was used to train teachers to use metacognition in their classrooms. Pre-service and in-service professional development programmes trained the participants to use metacognition in teaching by exposing the teachers to the various metacognitive pedagogies. The findings indicate that compared with control groups, the IMPROVE teachers outperformed their counterparts on pedagogical content knowledge (PCK), self-regulated learning (SRL), self-efficacy, judgment of learning (JOL), and their tendency to use in their classrooms what they have been taught in the course.

However, this area of research presents several limitations. Most of the reported studies were carried out on a relatively small scale. Moreover, the duration of most of these studies was short, about a month or semester. Only a few studies were carried out over a full academic year, or examined the lasting effects after one year. There is a need, therefore, to enlarge the scope of research to include large-scale as well as longitudinal studies across different age groups. This would require the development of additional assessment tools.

While all these studies focus on the effects of the metacognitive instructional methods on mathematics achievement, only a few assessed the effects on social-emotional outcomes such as motivation, maths anxiety or locus of control. Likewise, the examination of the effects of metacognitive guidance on judgment of learning and confidence of learning is still underdeveloped.

This book highlights the feasibility and effectiveness of the metacognitive approach which has been used in OECD countries including Belgium, Germany, the Netherlands, Israel, Singapore and the United States. Since the metacognitive interventions described here are generic methods, only small modifications would be required to use them in other disciplines. The application of IMPROVE to science education is a good example of how small modifications need to be and how effective they are in enhancing science literacy. International co-operation may facilitate these developments.

Curriculum and assessment policies can be key drivers of change. Often "you get what you assess" and "what is assessed is what is taught", not the other way around. The emphasis on mathematics and science literacy in international assessments of learning outcomes such as the OECD Programme for International Student Assessment (PISA) has started to change the teaching of these subjects, but mainly by individual teachers. This is not enough. Analyses of the curricula and textbooks across countries suggest that CUN tasks are rarely introduced in classrooms, and professional development seldom includes metacognitive interventions. The Singapore mathematics curriculum is an exception, exemplifying the inclusion of cognitive, metacognitive and affective components in the entire mathematics curriculum. IMPROVE and other metacognitive pedagogies highlight how schools could change their approaches to promote the teaching of CUN tasks and the activation of metacognition. Living in the global village calls for joint efforts in redesigning the mathematics (and science) curricula, textbooks, teaching methods and assessments.

Finally, educational systems have experienced tremendous changes in the last decades (OECD, 2014). The changes in student populations, teachers' professionalisation and school structures call for similar innovation in teaching methods. Dumont, Istance and Benavides (2010) pinpoint that "in a world where increasingly policy and practice are meant to be evidence-based, there is a need to take much more seriously the evidence on the nature of learning" (p. 335). Evidence-based practices provide the knowledge and mechanisms so as to move away from making decisions based on intuition; individual teachers do not have to "reinvent the wheel" by developing instructional methods. Instead, being well informed about what works and what does not work in education in general and in maths education in particular, could make change happen.

One approach for enhancing and sustaining learning in innovation-driven society lies in the implementation of metacognitive pedagogies that are based on solid theories about the nature of learning and provide a wealth of evidence on its impact on schooling outcomes. The fruitful links between research, practice, and policy described in this volume highlight the need for the continuing development of effective evidence-based teaching methods, and research into their benefits and trade-offs.

References

Azevedo, R. and V. Aleven (eds.) (2013), *International Handbook of Metacognition and Learning Technologies*, Springer, New York.

Dignath, C. and G. Buettner (2008), "Components of fostering self-regulated learning among students: A meta-analysis on intervention studies at primary and secondary school level", *Metacognition and Learning*, Vol. 3(3), pp. 231-264.

Dignath, C., G. Buettner, and H.P. Langfeldt (2008), "How can primary school students learn self-regulated learning strategies most effectively? A meta-analysis on self-regulation training programmes", *Educational Research Review*, Vol. 3(2), pp. 101-129.

Dumont, H., D. Istance and F. Benavides (eds.) (2010), The Nature of Learning: Using Research to Inspire Practice, *Educational Research and Innovation*, OECD Publishing, Paris, *http://dx.doi.org/10.1787/9789264086487-en*.

Hinton, C. and K.W. Fischer (2010), "Learning from the developmental and biological perspective", in H. Dumont, D. Istance and F. Benavides (eds.), *The Nature of Learning: Using Research to Inspire Practice*, Educational Research and Innovation, OECD Publishing, Paris, *http://dx.doi.org/10.1787/9789264086487-5-en*.

Lester, F.K., J.Garofalo and D.L. Kroll (1989), *The Role of Metacognition in Mathematical Problem Solving: A Study of Two Grade Seven Classes: Final Report*, ERIC Document Reproduction Service No. ED 314255.

Mevarech, Z. R., and Kramarski, B. (1997), "IMPROVE: A multidimensional method for teaching mathematics in heterogeneous classrooms", *American Educational Research Journal*, Vol. 34(2), pp. 365-395.

OECD (2014), *Measuring Innovation in Education: A New Perspective*, OECD Publishing, Paris, *http://dx.doi.org/ 10.1787/9789264215696-en*.

OECD (2008a), *Innovating to Learn, Learning to Innovate*, OECD Publishing, Paris, *http://dx.doi.org/10.1787/9789264047983-en*.

OECD (2008b), *Digital Learning Resources as Systemic Innovation in the Nordic Countries*, OECD publishing, Paris, *http://dx.doi.org/10.1787/9789264067813-en*.

OECD (2004), *The PISA 2003 Assessment Framework: Mathematics, Reading, Science and Problem Solving Knowledge and Skills, Education and Skills*, PISA, OECD Publishing, Paris, *http://dx.doi.org/10.1787/9789264101739-en*.

Schoenfeld, A.H. (1987), "What's all the fuss about metacognition?", in A.H. Schoenfeld (ed.), *Cognitive Science and Mathematics Education*, Lawrence Erlbaum, Hillsdale, NJ, pp. 189-215.

Slavin, R.E. (2010), "Co-operative learning: What makes group-work work?", in H. Dumont, D. Istance and F. Benavides (eds.), *The Nature of Learning: Using Research to Inspire Practice*, Educational Research and Innovation, OECD Publishing, Paris, *http://dx.doi.org/10.1787/9789264086487-9-en*.

Verschaffel, L., B. Greer and E. De Corte (2000), *Making Sense of Word Problems*, Swets and Zeitlinger, Lisse.

ORGANISATION FOR ECONOMIC CO-OPERATION AND DEVELOPMENT

The OECD is a unique forum where governments work together to address the economic, social and environmental challenges of globalisation. The OECD is also at the forefront of efforts to understand and to help governments respond to new developments and concerns, such as corporate governance, the information economy and the challenges of an ageing population. The Organisation provides a setting where governments can compare policy experiences, seek answers to common problems, identify good practice and work to co-ordinate domestic and international policies.

The OECD member countries are: Australia, Austria, Belgium, Canada, Chile, the Czech Republic, Denmark, Estonia, Finland, France, Germany, Greece, Hungary, Iceland, Ireland, Israel, Italy, Japan, Korea, Luxembourg, Mexico, the Netherlands, New Zealand, Norway, Poland, Portugal, the Slovak Republic, Slovenia, Spain, Sweden, Switzerland, Turkey, the United Kingdom and the United States. The European Union takes part in the work of the OECD.

OECD Publishing disseminates widely the results of the Organisation's statistics gathering and research on economic, social and environmental issues, as well as the conventions, guidelines and standards agreed by its members.

OECD PUBLISHING, 2, rue André-Pascal, 75775 PARIS CEDEX 16
(96 2014 02 1P) ISBN 978-92-64-21138-4 – 2014

Printed in Germany
by Amazon Distribution
GmbH, Leipzig